JUL. 1998

**American Herbal Products Association.**
**American Herbal Products Association's botanical safety**

AMERICAN HERBAL PRODUCTS ASSOCIATION'S
# BOTANICAL SAFETY HANDBOOK

# AMERICAN HERBAL PRODUCTS ASSOCIATION'S
# BOTANICAL SAFETY HANDBOOK

*Edited by*

MICHAEL MCGUFFIN, *Managing Editor*
CHRISTOPHER HOBBS, *L.Ac.*
ROY UPTON, *Herbalist*
ALICIA GOLDBERG

Prepared for the Standards Committee of the
American Herbal Products Association.

**CRC Press**
Boca Raton   Boston   London   New York   Washington, D.C.

**Library of Congress Cataloging-in-Publication Data**

American Herbal Products Association.
    American Herbal Products Association's botanical safety handbook :
guidelines for the safe use and labeling for herbs in commerce /
editors Michael McGuffin . . . [et al.] ; prepared for the Standards
Committee of the American Herbal Products Association.
       p.  cm.
    Includes bibliographical references (p.  ) and index.
    ISBN 0-8493-1675-8 (alk. paper)
    1. Herbs - - Toxicology.  2. Materia medica, Vegetable - - Toxicology.
    I. McGuffin, Michael.  II. Title.
    RA1250.A44  1997
    615'.321—dc21                                                         97-17162
                                                                                            CIP

This book contains information obtained from authentic and highly regarded sources. Reprinted material is quoted with permission, and sources are indicated. A wide variety of references are listed. Reasonable efforts have been made to publish reliable data and information, but the author and the publisher cannot assume responsibility for the validity of all materials or for the consequences of their use.

Neither this book nor any part may be reproduced or transmitted in any form or by any means, electronic or mechanical, including photocopying, microfilming, and recording, or by any information storage or retrieval system, without prior permission in writing from the publisher.

The consent of CRC Press LLC does not extend to copying for general distribution, for promotion, for creating new works, or for resale. Specific permission must be obtained in writing from CRC Press LLC for such copying.

Direct all inquiries to CRC Press LLC, 2000 Corporate Blvd., N.W., Boca Raton, Florida 33431.

© 1997 by CRC Press LLC

No claim to original U.S. Government works
International Standard Book Number 0-8493-1675-8
Library of Congress Card Number 97-17162
Printed in the United States of America    2 3 4 5 6 7 8 9 0
Printed on acid-free paper

# Preface

Increased attention on herbal products, both in the marketplace and in the legislative arena, has created a need for wider public access to data regarding the safety of botanicals. The passage of the Dietary Supplement Health and Education Act in October, 1994, furthered the need for such information, as this law authorizes the use of cautionary labeling for dietary supplements, including those that contain herbs.

The American Herbal Products Association (AHPA), through its Standards Committee, convened a special SubCommittee (hereinafter "the Committee") to address this need. The Committee members identified considerable safety data in varied texts and journals and discovered that some attempts to classify herbs had been undertaken in several other countries. No comprehensive compilation or review of this data for botanical ingredients sold in the North American marketplace, however, was available in a useful format.

The goal of the present work is to find a rational platform for the evaluation of herb safety, neither assuming that all natural substances are inherently safe, as some popular references suggest, nor blindly accepting reports of toxicity from uncritical sources. In undertaking this task the Editors met with information that presented significant challenges. Many authors utilize unreferenced data, perpetuate historical inaccuracies or display inherent biases against the use of botanicals. Also, contemporary reviews of the toxicity of many herbs are not available. Nonetheless, the Editors are confident that the body of information presented here is largely accurate. It is our sincere hope that readers of this work will find it to be a valuable reference and will address all useful criticisms to our attention.

In sponsoring this effort, the American Herbal Products Association addresses the common interest of industry, the public, and regulatory agencies in assuring safe access to a wide range of herbs and herbal products. This document provides accurate data to guide manufacturers and consumers in safe utilization of herbal products. As the most broadly established trade association in the herbal marketplace, AHPA has, by supporting and sponsoring the creation of this work, furthered the herb industry's leadership role in promoting the responsible use of herbs.

# Acknowledgements

The Committee and all of AHPA's members extend special thanks to Daniel Gagnon, whose vision inspired us to create this work. We also thank John Hallagan, Esq., of the Flavor and Extracts Manufacturing Association (FEMA), and William Appler for their encouragement, which solidified our own sense of value in this project. Beth Baugh selflessly provided warm and loving care of us during our long hours of work, for which we are all grateful. Beth also offered significant editorial insight that greatly enhanced the orderliness and readability of the final draft.

The development of the **Appendices** was overseen by Christopher Hobbs, L.Ac., and Michael Amster. Informed contributions to the writing and production of some of these sections were provided by a number of individuals, including: Chanchal Cabrera, M.N.I.M.H.; Bruce Canvasser, N.D.; David Hoffman, M.N.I.M.H.; Tori Hudson, N.D.; Amanda McQuade Crawford, M.N.I.M.H.; and Lisa Messerole, N.D. Additional assistance for this section was provided by Candy Carnes and Francis Fellows.

Valuable guidance was provided by each of our reviewers, and we extend thanks to: Dennis Awang, Ph.D.; Bruce Canvasser, N.D.; Beth Davis, Herbalist; Daniel Gagnon, Herbalist; Feather Jones, Herbalist; Ed Smith, Herbalist; John Staba, Ph.D.; Leslie Stevens, Herbalist; and David Winston, Herbalist.

Layout and design of the final manuscript was accomplished by Layne Jackson, whose company, GLOBEarts, is located in Chicago. In addition to her aesthetic sensibility, Layne exhibited patient toleration of our many last minute edits.

The Editors of this document provided their time and expertise with no financial remuneration and have donated all proceeds generated by its sale to AHPA. Thanks are therefore due to the following AHPA member companies who supported the Editors in the completion of this endeavor:

*Planetary Formulas, Rainbow Light Nutritional Systems,* and *Zand Herbal Formulas*.

This work was made possible by the following AHPA member company's generous financial sponsorship, which was offered without editorial consideration:

## Nutraceutical Corporation

Additional financial support was provided by AHPA members **Celestial Seasonings, Nature's Herbs** and **Botanicals International.**

# Introduction

## Determination of Herb Safety

In undertaking the tasks associated with the development of this document, the voices and experience of various organizations and individuals were considered. A primary source of guidance and inspiration has been the ongoing work of the World Health Organization (WHO).

In 1991, WHO's Programme on Traditional Medicines presented *Guidelines for the Assessment of Herbal Medicines* at the Sixth International Conference of Drug Regulatory Authorities. These guidelines, which were subsequently reviewed and adopted by WHO, propose that the safety of herbal medicine be assessed according to the following principle:

**"...that if the product has been traditionally used without demonstrated harm no specific restrictive regulatory action should be undertaken unless new evidence demands a revised risk-benefit assessment"** (Olayiwola, 1992).

The Committee has adopted this principle from the WHO *Guidelines* in the development of the present work. In his classic text *The Problem of Poisonous Plants*, J.M. Kingsbury provides further direction by noting that though there are many instances in which a plant contains a measurable amount of a toxic substance, the plant may be poisonous only if consumed in excessive quantities. He notes:

**"In order for a plant to be functionally poisonous, it must not only contain a toxic secondary compound, but also possess effective means of presenting that compound to an animal in sufficient concentrations, and the compound must be capable of overcoming whatever physiological or biochemical defenses the animal may possess against it. Thus the presence of a known poison principle, even in toxicologically significant amounts, in a plant does not automatically mean that either man or a given species of animal will ever be effectively poisoned by the plant"** (Kingsbury, 1979).

In examining the relevance of Kingsbury's position, it is of interest to revisit the means by which concerns for safety arise. Toxicity studies are often designed by feeding abnormally high quantities of an herb or isolated constituents of an herb to laboratory animals. For example, Bensky and Gamble report in their monograph on mulberry leaves that "Long-term use of 250 times the normal human dose in mice produced both liver and kidney damage" (Bensky and Gamble, 1986). Data based on excessive consumption have little relevance to the practical use of herbal supplements, and such findings are clearly not pertinent to normal human consumption patterns, any more than they would be for a food.

Significant toxicity data exist for isolated constituents of a wide variety of commonly available foods, as well as herbs. Potatoes, as a member of the Solanaceae family, contain trace amounts of the toxic glycoalkaloid solanine (Turner and Szczawinski, 1991). Although the symptoms of solanine poisoning are serious, potatoes themselves are generally considered to be a safe food. While consumption of as little as five grams of nutmeg can cause marked hallucinations (List and Horhämmer, 1973-79), no safety concern needs prevent us from enjoying a sprinkle of this characteristic flavor on Aunt Florence's eggnog. Similarly, no safety concern is associated with a candy flavored with peppermint oil, though as many as 26 toxins are reported to have been observed in the plant (Duke, 1989). Safety concerns for herbal products need not be extrapolated from constituent profiles with any more alarm than is appropriate for foods.

In following the principles espoused by WHO and incorporating the ideas delineated by Kingsbury, *it is imperative that herb safety be assessed according to the intended use of the substance within the historical context of its use.* The Committee intentionally refrained from extrapolating the toxicity of isolated constituents or considering excessive or irresponsible consumption patterns. The decision to place an herb in a restricted **Class** was made only if the use of the herb in a normal dosage range is documented as presenting a safety concern, or if the amount of a harmful or potentially harmful constituent obtainable from the crude plant is of sufficient quantity to be problematic. In all other cases in which the references report high dose or constituent related toxicity, an **Editors' Note** is appended for clarification.

## The Review Process and Classification

The Committee cross-referenced a thorough selection of standard toxicological and botanical references with a broad listing of herbs in trade. *Herbs of Commerce* (Foster, 1992) was used as the primary source for identifying herbs currently in the marketplace, though many additional herbs are included.

Classifications are included for each part of the plant with a history of use and are for dehydrated plant material, unless otherwise stated. Classifications address only the safety of the identified part of the herb in its whole, cut, or powdered form; as an ingredient in a finished product (tablets, capsules, teas, etc.); or as a crude extract to which no chemically-defined active substances have been added. Classifications are generally based on data which are associated with the use of the specific herb in isolation and in the quantities generally consumed for a therapeutic effect. Any cautions may therefore be somewhat overstated for an herb which appears in the market as part of a combination product, or for herbs which are used as foods and spices in less than such therapeutic quantities.

**Editors' Notes** and **Notices**, when included in a listing, are essential elements of overall classification. **Editors' Notes** include supplemental information relevant to the safe use of an herb, such as specific labeling recommendations, information regarding preparation, dose limits, possible adulteration, etc. In some cases the Committee found data which contradict the **Class** determined for a particular herb. The dissenting position is recorded in an **Editors' Note** along with the Committee's reasoning for discounting such data.

**Notices** were developed as a means for drawing attention to certain health conditions and plant constituents. A thorough discussion of each **Notice** is included in the *Appendix*.

Each herb is placed in a defined **Class** based on all the information included in the **Primary References**, any other data deemed as significant, and the experience of the Committee members. Central to the appropriate application of this document is the understanding that *classifications are based on an assumption of rational, informed use of herbs and herbal products*. **Classes** are defined as follows:

**Class 1**   Herbs which can be safely consumed when used appropriately.

**Class 2**   Herbs for which the following use restrictions apply, unless otherwise directed by an expert qualified in the use of the described substance:
  **2a:** For external use only.
  **2b:** Not to be used during pregnancy.
  **2c:** Not to be used while nursing.
  **2d:** Other specific use restrictions as noted.

**Class 3**   Herbs for which significant data exist to recommend the following labeling:
  "To be used only under the supervision of an expert qualified in the appropriate use of this substance." Labeling must include proper use information: dosage, contraindications, potential adverse effects and drug interactions, and any other relevant information related to the safe use of the substance.

**Class 4**   Herbs for which insufficient data are available for classification.

# Limitations of Scope

This work specifically excludes the following data, conditions, and related products:

- **Excessive consumption.** A classification implying safe use of an herb listed here does not warrant its safety in any quantity. Also, any toxicity concern which is significant only in excessive or immoderate use is not relevant to assignation of **Class**, though it may be referred to in an **Editors' Note**.

- **Safety or toxicity concerns based on isolated constituents.** As is the case with many common foods, some herbs are known to contain constituents which, in isolation, exhibit toxicity. Data based solely on constituents are not considered relevant to classification except in those cases where such compounds are known to accumulate, or where consumption patterns are sufficient to provide cause for health concerns. The presence of a constituent has been identified in a **Notice** if knowledge of the constituent is relevant to the safe use of an herb.

- **Toxicity data based solely upon intravenous or intraperitoneal administration.** The majority of products manufactured by members of AHPA are intended to be consumed orally and with adequate dosage instructions. Information associated with other forms of administration was reviewed but was not considered as a sole basis for classification. All classifications should be assumed to address oral administration, unless otherwise stated.

- **Traditional Chinese and Ayurvedic contraindications.** In Chinese and Ayurvedic therapeutic traditions, most herbs have contraindications based on an individual's constitutional strengths and weaknesses. These traditional concerns have not been included unless ignoring them would be harmful to a consumer.

- **Gastrointestinal disturbances.** Neither reports of nausea or emesis from excessive doses, nor occasional and/or minor gastrointestinal disturbances have been considered in establishing classification, unless frequency or severity of such reactions warrants consideration.

- **Potential drug interactions.** In any case in which the references establish a concern in regard to the interaction of an herb with medicinal substances, such concern has been noted. Since many of the substances listed here have not been thoroughly studied in this regard, it may be prudent to disclose consumption of herbs or herbal products when under the care of a health care provider.

- **Idiosyncratic reactions.** Any plant substance, whether used as food or medicine, has the potential to stimulate an unpredictable negative response in sensitive individuals. Classifications do not take into account such idiosyncratic responses.

- **Allergic reactions.** Certain plants in the Asteraceae, Apiaceae, and other plant families possess a relatively high degree of allergenicity. A plant's allergic potential is not considered a basis for restrictive classification. However, persons with a known allergy to ragweeds are advised to observe caution in the consumption of all plants of the Asteraceae family, especially the flowers.

- **Contact dermatitis.** The primary focus of this work is on herbal products for oral ingestion. Except in cases where there is a history of external therapeutic use, coupled with a record of associated dermatitis (e.g., mustard plasters), such concerns are beyond the scope of this document.

- **Well-known toxic plants which are not found in trade.** Many of the plants which are listed in standard toxicological texts as highly poisonous are not included in this document. Although isolates and constituents of some of these might be included in prescription drugs, they are not found in products which are otherwise accessible in a retail setting. Among the plants excluded are *Aconitum columbianum*, *Aconitum napellus*, *Adonis vernalis*, *Claviceps purpurea*, *Cantharanthus roseus*, *Chondrodendron tomentosum*, *Colchicum autumnale*, *Conium maculatum*, *Croton tiglium*, *Datura* spp., *Gelsemium sempervirens*, *Hyoscyamus niger*, *Nicotiana* spp., *Rauwolfia* spp., *Stramonium* spp., *Strophanthus kombe*, and *Strychnos nux-vomica*.

- **Homeopathic herbal preparations.** Such products are classified as over-the-counter or prescription drugs and are currently regulated by the *Homeopathic Pharmacopoeia of the United States*. This document does not address herbal products in homeopathic forms.

- **Essential oils.** Essential oils represent concentrations of specific volatile compounds. While many essential oils have a well-documented history of safe use by appropriately skilled persons, they often present toxicological concerns which are absent or moderate in the crude plant materials from which the oils are derived. The classification of essential oils is beyond the scope of this document.

- **Herbal products to which chemically-defined active substances, including chemically-defined isolated constituents of an herb, have been added.** Safety of such products should be determined by manufacturers and marketers prior to market introduction.

- **Environmental factors, additives, or contaminants.** Classifications do not consider potential adulteration of botanical materials, although known adulterants which present health risks are listed in an **Editor's Note**. Safety concerns can also arise due to environmental pollutants, pesticide residues, microbial contamination, or inappropriate storage conditions. Such concerns must be addressed by the manufacturing practices of suppliers and manufacturers who are responsible for assuring the purity of raw materials and are not considered in these classifications.

# References

In assembling the references from which to compile the data necessary for this effort, various criteria were considered. Contemporary texts concerned with herbs as they are used today in China, India, Europe, Australia, and the United States have been included. Additional information is provided by references which address the regulatory status, both actual and proposed, in several countries, including Germany and Britain, as well as in the U.S. and Canada. Several references focus solely on toxicity, with information included on consumption of plants by humans and other animals.

All of the toxicity data in each of the twenty-nine **Primary References**, listed and numbered at the beginning of the *Bibliography*, are included in this review. Such data, if they do not contribute to the specified **Class**, or if they are contradicted by other references, are included in the **Editors' Notes**.

Finally, one reference (#30) was reserved to represent consensus of the Committee, when relevant, as its members have significant experience in the therapeutic use and administration of many of the substances studied. This experience allowed the Editors to comment on some safety concerns not available in the published literature.

# Organization of Data

Listings are alphabetically arranged by Latin binomial. More than one species of a genus are combined into a single listing in those cases where two or more species are used interchangeably, or where the issues relevant to safe use are the same or nearly the same for related species. Following the Latin name is the botanical family name. In those few instances where two Latin binomials are used interchangeably, the less preferred of these is listed as a **Synonym**.

It is not unusual for a plant to have many common names, a fact which can confound the understanding of an herb's uses and potential safety concerns. AHPA published *Herbs of Commerce* (Foster, 1992) to address this concern, by assigning a single common name to each herb and by establishing a standard reference for common name nomenclature.

Common names in the present work are listed on either one or two lines. Entries are generally assigned a **Common Name** consistent with *Herbs of Commerce*. For plants not listed in *Herbs of Commerce* the common name which is most often cited in contemporary references is used. The Chinese (*pinyin*) name is included in italics for any herb which is primarily used in Traditional Chinese Medicine. Additional familiar common names are listed as **Other Common Names**.

Following the plant's names is the **Part** of the plant which is most often encountered in trade and for which the safety classification which follows is made. Occasional specific information is included for those herbs which require special processing. Also, some herbs supply more than one useful part. These parts are listed and classified together only in those cases where the safety issues of all parts are sufficiently similar; otherwise, separate listings are included for each plant part.

The remainder of each listing provides the key safety classification details which form the central purpose of this work. The first listed is **Class**. Each entry is assigned one or more of the classes described earlier in this introduction. Except for **Class 2d**, all class numbers are immediately followed by the references for such classification. A classification in **Class 2d** requires a statement which specifies the relevant concerns and concludes with the references for such statement.

Although in many cases the entry is complete at this stage, additional information is included in **Notices** and **Editors' Notes** for all plants where additional data are needed to thoroughly examine all safety issues. **Editors' Notes** and **Notices**, when included in a listing, are essential elements of overall classification.

There are fifteen **Notices** for constituents, such as caffeine or pyrrolizidine alkaloids, and twelve additional **Notices** for actions associated with particular plants, such as emetics and nervous system stimulants. The references which support a plant's inclusion in each **Notice** are included in brackets. In most cases, an associated **Class** and cautionary statement (usually following classification in **Class 2d**) are established for each **Notice**. A thorough discussion of each subject for which a **Notice** is used is included in either *Appendix 1* or *Appendix 2*.

**Editors' Notes** include supplemental information relevant to the safe use of an herb, such as specific labeling recommendations, information regarding preparation, dose limits, possible adulteration, etc. In some cases the Committee found data which contradict the **Class** determined for a particular herb. The dissenting position is recorded in an **Editors' Note** along with the Committee's reasoning for discounting such data.

**Standard Dose** is included only for those plants which list a recommendation that excessive dosage be avoided. The dose is usually given only in the quantity and form for direct consumption or for preparation as a tea and is based on the herb in its dried (dehydrated) form unless otherwise stated. Equivalent dosage in the form of tinctures and extracts must be calculated based on the concentration of such extracts on a dry weight basis. **Standard Dose** should not be taken to be the equivalent of a dosage limitation. Rather, this dosage should be seen as related to the concept of "serving size". Although **Standard Dose** may be relevant to the determination of appropriate dosage limits, a thorough examination of other specific factors would be required prior to setting such levels.

The following table provides a description of the organization of data. On occasion an additional Latin binomial is listed as a **Synonym**, immediately following the first listed Latin binomial.

| | |
|---|---|
| *Chelidonium majus* L.  [Papaveraceae] | -Latin binomial with authority and plant family. |
| **Common name:** celandine, *bai qu cai* | -Standard common name and *pinyin* name. |
| **Other common names:** greater celandine | -Other common names by which the plant is known. |
| **Part:** herb | -The part of the plant which is the subject of the classification. |
| **Class: 2b** [30]; **2d** - Not to be used by children [24]. | -The actual classification(s), with supporting references and appropriate cautions. |
| **Notice:** Berberine [28, (Duke, 1992), (Osol & Farrar, 1955)] *See page 136.* | -Additional relevant factors with references; a thorough discussion of each Notice is provided in either *Appendix 1* or *Appendix 2*. |
| **Ed. Note:** Although severe irritation of the digestive system is recorded in association with consumption of the *fresh* plant [15, 16, 18], the herb in its dried form is approved for internal use in Germany with no contraindications or side effects noted [4]. Frone & Pfänder [11] note that "In animal experiments, however, no such symptoms and no pathological changes in the digestive organs have been seen...More than 500 grams of greater celandine is said to be required to cause toxic effects in horses and cattle." Canadian regulations do not allow in foods [3]. An admonition against exceeding the stated dosage has been proposed in Australia, and the following labeling has been suggested: "May affect glaucoma treatment" [1]. | -Additional specific safety information. |
| **Standard Dose:** 600.0 mg to 5.0 grams per day [4, 26, 27] | Quantitative information, as required by dosage related caution above. |

# Disclaimer

The AHPA Standards Committee and the Editors of the *Botanical Safety Handbook* have endeavored to ensure that the information contained in this document accurately represents contemporary knowledge on the relative safety of crude botanicals. Particular care was given to the choice of primary references so as to provide meaningful cross-verification of the data on which classifications were made. However, the relative safety of any substance depends a great deal on the health of the consumer of the substance. Also, idiosyncratic of allergic reactions are often unpredictable. Any person or entity who utilizes an herb based on these classifications does so at their own risk and should consult a health care provider in the event of an adverse response.

There is no obligation at this time for AHPA members to adopt the information contained here in product labeling. Rather, this document is presented as a guideline, providing data to assist member and non-member manufacturers in developing labels that fully inform consumers. Verification of all data and classifications for the purpose of label development is the responsibility of the manufacturer.

# Contents

ACKNOWLEDGEMENTS
PREFACE
INTRODUCTION
SAFETY MONOGRAPHS, ALPHABETIZED BY LATIN BINOMIAL    1
APPENDIX 1: HERBAL CONSTITUENT PROFILES    129
    Aristolochic Acid    131
    Atropine    133
    β-asarone    134
    Berberine    136
    Cardiac Glycosides    138
    Cyanogenic Glycosides    141
    Estragole    143
    Iodine    145
    Lectins    146
    Oxalates    148
    Pyrrolizidine Alkaloids    149
    Safrole    152
    Salicylates    154
    Tannins    155
    Thujone    158
APPENDIX 2: HERBAL ACTION PROFILES    161
    Abortifacients    163
    Bulk-forming Laxatives    165
    Emetics    167
    Emmenagogues/Uterine Stimulants    169
    GI Irritants    172
    MAO Interaction    173
    Nervous System Stimulants    174
    Photosensitizing    176
    Stimulant Laxatives    177
APPENDIX 3: HERB LISTINGS BY CLASSIFICATION    181
    Herbs for external use only (class 2a)    182
    Herbs not to be used in pregnancy (class 2b)    183
    Herbs not to be used while breastfeeding (class 2c)    188
    Herbs not to be used without expert supervision (class 3)    189
PRIMARY REFERENCES    191
BIBLIOGRAPHY    199
PLANT INDEX    213

# BOTANICAL SAFETY HANDBOOK

## *Abies balsamea* (L.) Mill.   Pinaceae
**Common Name:** Canada balsam
**Other Common Names:** balm of Gilead, balsam fir
**Part:** bark
**Class:** 1

## *Achillea millefolium* L.   Asteraceae
**Common Name:** yarrow
**Other Common Names:** milfoil
**Class:** 2b [18, 30]
**Notice:** Emmenagogue/Uterine Stimulant [30] *See page 169.*
**Ed. Note:** With external use, Wichtl [27] advises immediate cessation of treatment in the event of "itching and inflammatory changes in the skin" for persons prone to allergies to Asteraceae, while Leung & Foster [17] contraindicate yarrow for those with such allergies.

Federal regulations in the U.S. require that finished food or beverage products containing yarrow be thujone-free [20], though the constituent is present in only trace amounts [17].

The toxicity concerns cited in Kingsbury [15] and attributed to Pammel (Pammel, 1910) are not relevant to human consumption patterns.

## *Achyranthes bidentata* Bl.   Amaranthaceae
**Common Name:** achyranthes, *nui xi*
**Part:** root
**Class:** 2b [2]; 2d - Contraindicated in excessive menstruation [2].
**Notice:** Emmenagogue/Uterine Stimulant [2] *See page 169.*
**Ed. Note:** Various species can be traded under the name *nui xi*. *A. longifolia* is also contraindicated in pregnancy [23].

## *Aconitum carmichaelii* Debx.   Ranunculaceae
**Common Name:** aconite, *fu zi*
**Part:** prepared root
**Class:** 3 [1, 2, 13, 17, 23, 29]
**Ed. Note:** Other species of *Aconitum* are also in trade [13, 14]. All should be considered Class 3, whether prepared or in an untreated state. Even external application is reported to cause toxic symptoms (Mitchell et al., 1983).

## *Acorus calamus* L.   Araceae
**Common Name:** calamus
**Other Common Names:** acorus, sweet flag
**Part:** rhizome of the diploid asarone-free variety (*Acorus calamus* L. var. *americanus* Wolff).

Class: 1
Ed. Note: Since varieties of calamus may not be well differentiated in commerce, the cautions stated for the Asian and European varieties should be considered relevant to any sample which is not positively identified as the American variety.

All varieties of calamus are prohibited in foods in the U.S. [20]. Canadian regulations list A. *calamus* as an unacceptable non-medicinal ingredient for oral use products (Michols, 1995).

Part: rhizome of the asarone-containing triploid and tetraploid (Asian and European) varieties.
Class: 2b [30]; 3 [6, 8, 17, 21, 23, 27, 28]
Notice: β-asarone (usually 1.1-2.6%; up to 8.0%) [6, 8, 17, 21, 23, 27, 28] See page 134.

## *Acorus gramineus* Soland.      Araceae
Common Name: grass-leaved calamus, *shi chang pu*
Other Common Names: grassy-leaved sweetflag
Part: rhizome
Class: 2b [30]; 3 [1, 2, 23, 29, 30]
Notice: β-asarone, 0.08 - 0.3%) [7, 29] See page 134.
Ed. Note: Canadian regulations do not allow grass-leaved calamus in food, except to flavor alcoholic beverages if the material is asarone-free [3].

## *Adenophora stricta* Miq.      Campanulaceae
## *Adenophora tetraphylla* (Thunb.) Fisch.
Common Name: adenophora, *nan sha shen*
Part: root
Class: 1
Ed. Note: Although Huang [14] reports an unreferenced hemolytic effect on red blood cells, many saponin glycosides exhibit *in vitro* hemolytic activity which is not indicative of an *in vivo* safety concern.

## *Adiantum pedatum* L.      Adiantiaceae
Common Name: maidenhair fern
Part: herb
Class: 2b [6, 18]; 2d - Large doses are emetic [6, 18].
Ed. Note: Regulated in the U.S. as an allowable flavoring agent in alcoholic beverages only [20].

---

Class 1: Herbs that can be safely consumed when used appropriately.
Class 2: Herbs for which the following use restrictions apply, unless otherwise directed by an expert qualified in the use of the described substance:
    (2a) For external use only;      (2b) Not to be used during pregnancy;
    (2c) Not to be used while nursing;      (2d) Other specific use restrictions as noted.

*Agastache rugosa* (Fisch. et Mey.) O. Ktze.  **Lamiaceae**
Common Name: agastache, *tu huo xiang*
Part: herb
Class: 1


*Agrimonia eupatoria* L.  **Rosaceae**
Common Name: agrimony
Part: herb
Class: 1
Notice: Tannins (4.0-10.0%) [27] *See page 155.*


*Albizia julibrissin* Durazz.  **Fabaceae**
Common Name: silk tree, *he huan pi*
Other Common Names: mimosa tree
Part: bark
Class: 2b [30]
Notice: Tannins [2, 13, 29] *See page 155.*
Emmenagogue/Uterine Stimulant [13, 29] *See page 169.*

Part: flowers
Class: 1


*Alcea rosea* L.  **Malvaceae**
Common Name: hollyhock
Part: root
Class: 1


*Alchemilla xanthochlora* Rothm.  **Rosaceae**
Common Name: lady's mantle
Other Common Names: alder buckthorn
Part: herb
Class: 1


*Aletris farinosa* L.  **Liliaceae**
Common Name: aletris
Other Common Names: stargrass, true unicorn, starwort, blazing star
Part: rhizome and root
Class: 2d - Drug interaction: antagonizes some oxytocins (pitocin) [18].

---

Class 3: Herbs for which significant data exist to recommend the following labeling: "To be used only under the supervision of an expert qualified in the appropriate use of this substance." Labeling must include proper use information: dosage, contraindications, potential adverse effects and drug interactions, and any other relevant information related to the safe use of this substance.
Class 4: Herbs for which insufficient data are available for classification.

**Ed. Note:** According to Grieve, "...fresh root in large doses somewhat narcotic, emetic and cathartic; when dried these properties are lost" (Grieve, 1931). Solis-Cohen and Githens state, "In small doses, about 10 grains [±650 mg], it is a simple bitter; in larger doses it is cathartic and emetic" (Solis-Cohen and Githens, 1928).

Canadian regulations do not allow aletris as a non-medicinal ingredient for oral use products (Michols, 1995).

### *Alisma orientale* (Sam.) Juzepczuk                Alismataceae
**Common Name:** alisma, *ze xie*
**Part:** rhizome
**Class: 2d** - Prolonged use may cause gastrointestinal irritation [2].
**Ed. Note:** Renal and hepatotoxicity are associated with doses 20 to 40 times effective clinical dose, administered to rats for 90 days [7].

### *Alkanna tinctoria* (L.) Tausch                Boraginaceae
**Common Name:** alkanet
**Part:** root
**Class: 2a** [8, 24, 27, 30]; **2b** [8]; **2c** [8]; **2d**- long-term use is not recommended [30].
**Notice:** Pyrrolizidine Alkaloids [8, 18, 24, 27, 28] *See page 149.*
**Ed. Note:** The AHPA Board of Trustees recommends that all products with botanical ingredient(s) which contain toxic pyrrolizidine alkaloids, including *Alkanna tinctoria*, display the following cautionary statement on the label:

"For external use only. Do not apply to broken or abraded skin. Do not use when nursing."

### *Allium sativum* L.                Liliaceae
**Common Name:** garlic
**Part:** bulb
**Class: 2c** [6, 30]
**Ed. Note:** The classifications and concerns for this herb are based upon therapeutic use and dosage and may not be relevant to its consumption as a spice.

Occasional isolated side effects are associated with consumption of garlic: gastrointestinal disturbance in sensitive individuals is reported by several references [4, 5, 8, 10, 16]; a single case of platelet dysfunction in an 87-year-old patient was reported after chronic consumption of approximately 2 grams daily of garlic cloves [9]; hypoglycemic activity and an increase in serum insulin

---

Class 1: Herbs that can be safely consumed when used appropriately.
Class 2: Herbs for which the following use restrictions apply, unless otherwise directed by an expert qualified in the use of the described substance:
  (2a) For external use only;     (2b) Not to be used during pregnancy;
  (2c) Not to be used while nursing;   (2d) Other specific use restrictions as noted.

have been noted in mildly diabetic rabbits fed garlic [6]; Watt and Breyer-Brandwijk [25] state that "...oral administration [of fresh garlic] to children is said to be dangerous and even fatal." Also, Chadha [6] contraindicates garlic in pregnancy. However, the long history of use as a food has established that consumption of garlic in reasonable quantities is generally safe, though caution may be indicated in children and while breastfeeding.

The above classifications and notes may not apply to processed garlic products.

## *Allium schoenoprasum* L. — Liliaceae
**Common Name:** chives
**Part:** leaf
**Class:** 1

## *Aloe ferox* Mill. — Aloaceae
## *Aloe perryi* Baker
## *Aloe vera* (L.) N. L. Burm. (=A. *barbadensis* Mill.)
**Common Name:** aloe vera
**Other Common Names:** cape aloe
**Part:** mucilaginous leaf gel from parenchymatious leaf cells
**Class (Internal use):** 1
**Class (External use):** 2d - May delay wound healing following laparotomy or caesarean delivery (Tyler, 1994; Pizzorno & Murray, 1992).
**Ed. Note:** The leaf gel is commonly consumed in quantity as a cleansing juice preparation. It does not act as a strong cathartic like the dried juice from the pericyclic region of the leaves which is a distinct product from a different region of the leaves.
**Common Name:** aloes
**Part:** dried juice from the pericyclic region of the leaves
**Class:** 2b [2, 4, 5, 6, 22, 24, 26, 27, 29]; 2c [4, 5, 24, 27]; 2d - Contraindicated in intestinal obstruction, abdominal pain of unknown origin, or any inflammatory condition of the intestines (appendicitis, colitis, Crohn's disease, irritable bowel syndrome, etc.) [4, 5, 9, 10, 21]; in hemorrhoids [5, 10, 18, 22, 24]; in kidney dysfunction [24, 27]; in menstruation [24, 27]; and in children less than 12 years of age [4, 5, 9]; not for use in excess of 8-10 days [4, 5, 9, 17, 24, 26, 27].
**Standard Dose:** 50-300 mg in a single dose at bedtime [5, 28, (Federal Register, 1985)].
**Notice:** Stimulant Laxative [2, 4, 5, 6, 9, 10, 13, 17, 21, 22, 24, 26, 27, 28, 29]
*See page 177.*

Class 3: Herbs for which significant data exist to recommend the following labeling: "To be used only under the supervision of an expert qualified in the appropriate use of this substance." Labeling must include proper use information: dosage, contraindications, potential adverse effects and drug interactions, and any other relevant information related to the safe use of this substance.
Class 4: Herbs for which insufficient data are available for classification.

**Ed. Note:** Overdose (stated at as little as 1.0 gram per day for several days) can cause colonic perforation and bleeding diarrhea, as well as kidney damage, and according to one reference, death [5, 17, 18, 21, 27, 28, 29].

Aloe is generally considered to be obsolete as a laxative, as it is a less desirable agent for laxation than other available substances [21, (Hoover, 1970)].

AHPA recommends the following label for products containing this herb in sufficient quantity to warrant such labeling: Do not use this product if you have abdominal pain or diarrhea. Consult a health care provider prior to use if you are pregnant or nursing. Discontinue use in the event of diarrhea or watery stools. Do not exceed recommended dose. Not for long-term use.

### *Aloysia triphylla* (L'Her.) Britton — Verbenaceae
**Common Name:** lemon verbena
**Other common name:** *Lippia citriodora*
**Part:** leaf
**Class:** 1
**Ed. Note:** Although officially regulated in the U.S. as an allowable flavoring agent in alcoholic beverages only [20], this limitation appears to be unnecessarily cautious in light of its long history of use as a tea.

### *Alpinia galanga* (L.) Willd. — Zingiberaceae
**Common Name:** greater galangal
**Part:** rhizome
**Class:** 1
**Ed. Note:** Although classification in the U.S. allows this species to be used only as a flavoring agent in alcoholic beverages, the closely related and historically interchangeable *A. officinarum* is listed in the same reference as "generally recognized as safe" [20]. Based on the long history of use of both of these species in food and medicine, a reconsideration of the caution referenced here is in order.

### *Alpinia officinarum* Hance — Zingiberaceae
**Common Name:** lesser galangal
**Part:** rhizome
**Class:** 1

---

Class 1: Herbs that can be safely consumed when used appropriately.
Class 2: Herbs for which the following use restrictions apply, unless otherwise directed by an expert qualified in the use of the described substance:
    (2a) For external use only;    (2b) Not to be used during pregnancy;
    (2c) Not to be used while nursing;    (2d) Other specific use restrictions as noted.

*Althaea officinalis* L.  Malvaceae
**Common Name:** marshmallow
**Part:** root, leaf, and flower
**Class:** 1
**Ed. Note:** Commission E [4] notes without reference that "absorption of other drugs taken simultaneously may be delayed."


*Amomum melegueta* Roscoe  Zingiberaceae
**Common Name:** grains of paradise
**Part:** fruit and seed
**Class:** 1


*Amomum tsao-ko* Crev. et. Lem.  Zingiberaceae
**Common Name:** *cao guo*
**Part:** fruit
**Class:** 1


*Anaphalis margaritacea* (L.) Benth. & J. D. Hook.  Asteraceae
**Common Name:** pearly everlasting
**Part:** herb
**Class:** 1


*Andrographis paniculata* (Burm. f.) Nees  Acanthaceae
**Common Name:** andrographis, *chuan xin lian*
**Part:** herb
**Class:** 2b [7, 14]
**Notice:** Abortifacient [7, 14] *See page 163*.
**Ed. Note:** Large oral doses are reported to cause gastric discomfort and loss of appetite [7].


*Anemarrhena asphodeloides* Bge.  Liliaceae
**Common Name:** anemarrhena, *zhi mu*
**Part:** rhizome
**Class:** 1

---

Class 3: Herbs for which significant data exist to recommend the following labeling: "To be used only under the supervision of an expert qualified in the appropriate use of this substance." Labeling must include proper use information: dosage, contraindications, potential adverse effects and drug interactions, and any other relevant information related to the safe use of this substance.
Class 4: Herbs for which insufficient data are available for classification.

*Anemopsis californica* (Nutt.) Hook. & Arn.  Saururaceae
Common Name: yerba mansa
Part: rhizome and root
Class: 1

*Anethum graveolens* L.  Apiaceae
Common Name: dill
Part: herb and "fruit (commonly know as "seed")"
Class: 1

*Angelica archangelica* L.  Apiaceae
*Angelica atropurpurea* L.
Common Name: angelica
Part: root and fruit (commonly know as "seed")
Class: 2b [18]; 2d - Avoid prolonged exposure to sunlight [4, 28].
Notice: Emmenagogue/Uterine Stimulant [10] *See page 169.*
Photosensitizing [4, 28] *See page 176.*
Ed. Note: Leung & Foster [17] note that the coumarins responsible for the phototoxicity of the root and seed of A. *archangelica* are not found in the steam-distilled oils of these.

Canadian regulations list A. *archangelica* as an unacceptable non-medicinal ingredient for oral use products (Michols, 1995).

*Angelica dahurica* (Fischer ex Hoffm.) Franch. et Sav.  Apiaceae
Common Name: fragrant angelica, *bai zhi*
Part: root
Class: 1
Ed. Note: Other species of *Angelica* are commonly traded as *bai zhi* [2, 7, 13, 29].

*Angelica pubescens* Maxim.  Apiaceae
Common Name: pubescent angelica, *du huo*
Part: root
Class: 2d - Avoid prolonged exposure to sunlight [30].
Notice: Photosensitizing [7] *See page 176.*
Ed. Note: Other species of *Angelica* as well as plants of other genera are commonly traded as *du huo* [7, 13].

---

Class 1: Herbs that can be safely consumed when used appropriately.
Class 2: Herbs for which the following use restrictions apply, unless otherwise directed by an expert qualified in the use of the described substance:
  (2a) For external use only;      (2b) Not to be used during pregnancy;
  (2c) Not to be used while nursing;   (2d) Other specific use restrictions as noted.

## *Angelica sinensis* (Oliv.) Diels            Apiaceae

**Common Name:** dong quai, *dang gui, tang kuei*
**Part:** root
**Class:** 2b [3, 30]
**Ed. Note:** Both a stimulating and relaxing effect on the uterus have been reported for this species [13], but no contraindication during pregnancy is noted in any of the Chinese herbal references cited in this work [2, 3, 13, 14]. Canadian regulations, however, require bilingual label warnings against the use of dong quai during pregnancy [3] and do not allow it as a non-medicinal ingredient in oral use products (Michols, 1995).

## *Aniba rasaeodora* Ducke            Lauraceae

**Common Name:** Brazilian rosewood
**Part:** bark
**Class:** 1

## *Anthriscus cerefolium* (L.) Hoffm.            Apiaceae

**Common Name:** chervil
**Part:** herb
**Class:** 2b [6, 18]
**Ed. Note:** The classifications and concerns for this herb are based upon therapeutic use and dosage and may not be relevant to its consumption as a spice.

## *Apium graveolens* L.            Apiaceae

**Common Name:** celery
**Part:** fruit (commonly know as "seed")
**Class:** 2b [5]; 2d - Individuals with renal disorders should use with caution [5, 18, 24, 27].
**Notice:** Photosensitizing [30] *See page 176.*
**Ed. Note:** The classifications and concerns for this herb are based upon therapeutic use and dosage and may not be relevant to its consumption as a spice.
    Commission E [4] reports potential allergenicity, including anaphylactic shock.

---

Class 3: Herbs for which significant data exist to recommend the following labeling: "To be used only under the supervision of an expert qualified in the appropriate use of this substance." Labeling must include proper use information: dosage, contraindications, potential adverse effects and drug interactions, and any other relevant information related to the safe use of this substance.
Class 4: Herbs for which insufficient data are available for classification.

*Apocynum androsaemifolium* L.     **Apocynaceae**
*Apocynum cannabinum* L.
**Common Name:** spreading dogbane (*A. androsaemifolium*); Indian hemp (*A. cannabinum*).
**Other Common Names:** bitter root, milkweed, wallflower, wild ipecac (*A. androsaemifolium*), Indian physic (*A. cannabinum*).
**Part:** root
**Class:** 3 [1, 18]
**Notice:** Cardiac Glycosides [1, 18, 25, 28] *See page 138.*
**Ed. Note:** Canadian regulations do not allow spreading dogbane in foods [3].

*Aralia californica* Wats.     **Araliaceae**
**Common Name:** California spikenard
**Part:** rhizome, roots
**Class:** 2b [30]

*Aralia nudicaulis* L.     **Araliaceae**
**Common Name:** small spikenard
**Other Common Names:** false sarsaparilla, wild licorice
**Part:** rhizome
**Class:** 2b [30]

*Aralia racemosa* L.     **Araliaceae**
**Common Name:** spikenard
**Other Common Names:** spignet
**Part:** rhizome
**Class:** 2b [30]

*Arctium lappa* L.     **Asteraceae**
**Common Name:** burdock
**Part:** seed and root
**Class:** 1
**Ed. Note:** Although De Smet [8] and Wren [28] report a single poisoning with "atropine-like" symptoms, this isolated case has been shown to be due to adulteration with belladonna (Burnham, 1996). De Smet's later volume [9] clarifies this point by stating that the "principal risk appears to be...due to adulteration or contamination with belladonna root."

---

Class 1: Herbs that can be safely consumed when used appropriately.
Class 2: Herbs for which the following use restrictions apply, unless otherwise directed by an expert qualified in the use of the described substance:
   (2a) For external use only;     (2b) Not to be used during pregnancy;
   (2c) Not to be used while nursing;     (2d) Other specific use restrictions as noted.

## *Arctostaphylos uva-ursi* (L.) Spreng.  Ericaceae
**Common Name:** uva-ursi
**Other Common Names:** bearberry, kinnikinnik
**Part:** leaf
**Class:** 2b [3, 5, 24, 26]; 2d - Contraindicated in kidney disorders [5, 17], irritated digestive conditions [17, 27], and with acidic urine or in conjunction with remedies which produce acidic urine [4]; not for prolonged use without consulting a practitioner [4].
**Notice:** GI Irritant [4, 24, 26] *See page 172.*
Tannins (6.0-27.5%) [17, 27] *See page 155.*
**Ed. Note:** Canadian regulations do not allow uva-ursi as a non-medicinal ingredient for oral use products (Michols, 1995).
   Although *Buxus* leaf, a **Class 3** herb, is reported as an adulterant of uva-ursi [18], Wichtl [27] states that this is "...rarely found to occur in practice."

## *Arisaema japonicum* Blume  Araceae
**Common Name:** Japanese arisaema, *tian nan xing*
**Part:** processed tuber (Tu, 1988)
**Class:** 2b [2, 29]
**Ed. Note:** The fresh or unprocessed tuber is toxic [2, 7].

## *Arisaema triphyllum* (L.) Torr.  Araceae
**Common Name:** Jack-in-the-pulpit
**Part:** tuber
**Class:** 1
**Ed. Note:** Although the tuber has been classified as toxic [12, 19], toxicity is destroyed by proper drying [10, 19]. Muenscher [19] also states that large doses may cause gastroenteritis without clarifying if this report is relevant to fresh or dried material.

## *Aristolochia clematitis* L.  Aristolochiaceae
**Common Name:** long birthwort
**Part:** root, rhizome, herb
**Class:** 2b [23]; 3 [8, 18, 28]
**Notice:** Aristolochic acid (herb: 0.03-0.3%; root: 0.4-1.1%) [8, 18, 28] *See page 131.*
**Ed. Note:** All reported mutagenicity is based on animal studies with pure aristolochic acid. Weiss [26] argues that "Empiricism...contradicts the animal studies...", since there is a long history of use with no reports of carcinogenic

---

Class 3: Herbs for which significant data exist to recommend the following labeling: "To be used only under the supervision of an expert qualified in the appropriate use of this substance." Labeling must include proper use information: dosage, contraindications, potential adverse effects and drug interactions, and any other relevant information related to the safe use of this substance.
Class 4: Herbs for which insufficient data are available for classification.

effects. However, recent concerns regarding weight loss preparations which contain herbs high in aristolochic acid have raised the question of the potential for kidney damage in humans (Vanherweghem, 1994; Moffet, 1995).

Related species are potentially lethal to animals [25].

Regulated in the U.S. as an allowable flavoring agent in alcoholic beverages only [20].

### *Aristolochia contorta* Bge.     Aristolochiaceae
### *Aristolochia debilis* Sieb. et Zucc.
**Common Name:** *ma dou ling* (fruit); *qing mu xiang* (root); *tian xian teng* (vine)
**Part:** fruit, root, vine (herb)
**Class:** 2b [23]; 3 [7, 29]
**Notice:** Aristolochic acid [7] *See page 131.*
**Ed. Note:** See *A. clematitis.*

### *Aristolochia serpentaria* L.     Aristolochiaceae
**Common Name:** Virginia snakeroot
**Other Common Names:** serpentaria
**Part:** rhizome
**Class:** 2b [23]; 3 [1, 8, 18]
**Notice:** Aristolochic acid (root and rhizome 0.046%) [1, 8, 18] *See page 131.*
**Ed. Note:** See *A. clematitis*

### *Armoracia rusticana*     Brassicaceae
**Common Name:** horseradish
**Part:** rhizome and root
**Class:** 2d -Contraindicated with inflammation of the gastric mucosa [4, 18] and with kidney disorders [4]; not to be used by children under 4 years old [4].
**Ed. Note:** The classifications and concerns for this herb are based upon therapeutic use and may not be relevant to its consumption as a spice.

### *Arnica latifolia* Bong.     Asteraceae
### *Arnica montana* L.
### *Arnica* spp.
**Common Name:** arnica
**Part:** whole plant, flower, and rhizome
**Class, External Use:** 2d - Do not use on open wounds or broken skin [30].
**Ed. Note:** Can cause allergic dermatitis in sensitive persons or with prolonged use [4, 8, 10, 11, 17, 21, 27].

---

Class 1: Herbs that can be safely consumed when used appropriately.
Class 2: Herbs for which the following use restrictions apply, unless otherwise directed by an expert qualified in the use of the described substance:
    (2a) For external use only;     (2b) Not to be used during pregnancy;
    (2c) Not to be used while nursing;     (2d) Other specific use restrictions as noted.

**Class, Internal Use:** 2b [30]; 3 [8, 10, 17, 18, 24, 27]

**Ed. Note:** Regulated in the U.S. as an allowable flavoring agent in alcoholic beverages only [20]. Canadian regulations list *A. montana* as an unacceptable non-medicinal ingredient for oral use products (Michols, 1995).

## *Artemisia abrotanum* L.     Asteraceae
**Common Name:** southernwood
**Part:** herb
**Class:** 2b [30]
**Notice:** Emmenagogue/Uterine Stimulant [30] *See page 169.*

## *Artemisia absinthium* L.     Asteraceae
**Common Name:** wormwood
**Other Common Names:** absinthe
**Part:** herb
**Class:** 2b [30]; 2c [30]; 2d - Not for long-term use; do not exceed recommended dose [4, 17, 27].
**Standard Dose:** Not to exceed 1.5 grams of dried herb in tea, 2 to 3 times daily [4, 17, 27, 28].
**Notice:** Thujone (up to 0.6%) [17, 27] *See page 158.*
Emmenagogue/Uterine Stimulant [30] *See page 169.*
**Ed. Note:** The primary references do not substantiate toxicological concern, except with long-term use [17], excessive dose, or use of the essential oil [10, 26, 27].

Regulatory restrictions against the use of *A. absinthium* requiring finished food products to be thujone-free exist in the U.S. and other countries [17, 20, 21]. Wormwood was recently recommended for removal from Canada's list of food adulterants (Welsh, 1995).

## *Artemisia annua* L.     Asteraceae
**Common Name:** sweet annie, *qing hao*
**Part:** herb
**Class:** 2b [30]

## *Artemisia capillaris* Thunb.     Asteraceae
## *Artemisia scoparia* Waldst. et Kit.
**Common Name:** capillary artemisia (*A. capillaris*), *yin chen hao*
**Part:** herb
**Class:** 2b [30]

---

Class 3: Herbs for which significant data exist to recommend the following labeling: "To be used only under the supervision of an expert qualified in the appropriate use of this substance." Labeling must include proper use information: dosage, contraindications, potential adverse effects and drug interactions, and any other relevant information related to the safe use of this substance.
Class 4: Herbs for which insufficient data are available for classification.

**Ed. Note:** Adverse effects may include nausea, abdominal distention, and dizziness [7, 14]. Two women treated for infectious hepatitis with *A. capillaris* and *Zizyphus jujuba* developed Adams-Stokes syndrome [2, 7].

*Artemisia douglasiana* Bess.  Asteraceae
*Artemisia lactiflora* Wall. ex DC.
*Artemisia vulgaris* L.
**Common Name:** mugwort
**Part:** herb
**Class:** 2b [4, 6, 18, 28]
**Notice:** Emmenagogue/Uterine Stimulant [28] *See page 169.*

*Artemisia dracunculus* L. 'Sativa'  Asteraceae
**Common Name:** French tarragon
**Part:** herb
**Class:** 1
**Notice:** Estragole (81% of the essential oil) [27] *See page 143.*

*Asarum canadense* L.  Aristolochiaceae
**Common Name:** Canada snakeroot
**Other Common Names:** wild ginger
**Part:** rhizome
**Class:** 2b [23]; 2d - Not for long-term use; do not exceed recommended dose [30].
**Standard Dose:** 2.0-4.0 grams (Wood & Osol, 1943).
**Notice:** Emmenagogue/Uterine Stimulant [10] *See page 169.*
β-asarone [18] *See page 134.*
Aristolochic acid [18, 28] *See page 131.*
**Ed. Note:** The rhizome contains from 3.5-4.5% essential oil [28], with only traces of β-asarone [18].

*Asarum europaeum* L.  Aristolochiaceae
**Common Name:** asarabacca
**Part:** rhizome
**Class:** 2b [23]; 2d - Not for long-term use; do not exceed recommended dose [30]; may cause nausea and vomiting [4, 10, 28].
**Standard Dose:** 650-800 mg [10].
**Notice:** Emetic [4, 10, 28] *See page 167.*
β-asarone (0.35-2.0%) [8, 28] *See page 134.*

---

Class 1: Herbs that can be safely consumed when used appropriately.
Class 2: Herbs for which the following use restrictions apply, unless otherwise directed by an expert qualified in the use of the described substance:
    (2a) For external use only;    (2b) Not to be used during pregnancy;
    (2c) Not to be used while nursing;    (2d) Other specific use restrictions as noted.

## *Asclepias asperula* (Dcne.) Woodson     Asclepiadaceae
Common Name: inmortal
Part: root
Class: 2b (Moore, 1979)
Ed. Note: Many *Asclepias* species contain cardiac glycosides and toxic resinoids [18]. While cases of livestock poisoning associated with consumption of the above ground parts are recorded [15, 12], "...there are apparently no reported fatal poisonings of humans by milkweed" (Turner and Szazawinski, 1991).

## *Asclepias tuberosa* L.     Asclepiadaceae
Common Name: pleurisy root
Part: root
Class: 2b [18]; 2d - May cause nausea and vomiting [18].
Notice: Emetic [18] *See page 167.*
Cardiac Glycosides [18] *See page 138.*
Ed. Note: See *A. asperula.*
    Canadian regulations do not allow pleurisy root as a non-medicinal ingredient for oral use products (Michols, 1995).

## *Asparagus cochinchinensis* (Lour.) Merr.     Liliaceae
Common Name: Chinese asparagus, *tian men dong*
Part: processed rhizome (Tu, 1988)
Class: 1

## *Asparagus officinalis* L.     Liliaceae
Common Name: asparagus
Part: rhizome
Class: 2d - Contraindicated in inflammatory kidney disorders [4].
Ed. Note: Although Commission E [4] contraindicates this botanical in irrigation therapy when edema is caused by impaired kidney or heart function, this concern is not relevant to oral consumption.

## *Astragalus membranaceus* Bunge     Fabaceae
## *Astragalus mongholicus* Bunge.
Common Name: astragalus, *huang qi*
Part: root
Class: 1

---

Class 3: Herbs for which significant data exist to recommend the following labeling: "To be used only under the supervision of an expert qualified in the appropriate use of this substance." Labeling must include proper use information: dosage, contraindications, potential adverse effects and drug interactions, and any other relevant information related to the safe use of this substance.
Class 4: Herbs for which insufficient data are available for classification.

***Atractylodes lancea*** **(Thunb.) DC.**            **Asteraceae**
**Common Name:** southern tsangshu, *cang zhu*
**Part:** rhizome
**Class:** 1

***Atractylodes macrocephala*** **Koidz.**            **Asteraceae**
**Common Name:** atractylodes, *bai zhu*
**Part:** rhizome
**Class:** 1

***Atropa belladonna*** **L.**            **Solanaceae**
**Common Name:** belladonna
**Other Common Names:** deadly nightshade
**Part:** leaf
**Class:** 3 [1, 4-6, 8, 10-12, 15-22, 24-26, 28]
**Notice:** Atropine (0.3-0.6% total tropane alkaloids) [4, 5, 17, 21, 28] *See page 133.*
**Ed. Note:** Canadian regulations do not allow belladonna in foods [3].

***Avena fatua*** **L.**            **Poaceae**
**Common Name:** wild oat
**Part:** spikelets
**Class:** 1

***Avena sativa*** **L.**            **Poaceae**
**Common Name:** oats
**Part:** spikelets
**Class:** 1

***Ballota nigra*** **L.**            **Lamiaceae**
**Common Name:** black horehound
**Part:** herb
**Class:** 1

***Baptisia tinctoria*** **(L.) R.Br. ex Ait. f.**            **Fabaceae**
**Common Name:** wild indigo
**Part:** root

---

Class 1: Herbs that can be safely consumed when used appropriately.
Class 2: Herbs for which the following use restrictions apply, unless otherwise directed by an expert qualified in the use of the described substance:
    (2a) For external use only;      (2b) Not to be used during pregnancy;
    (2c) Not to be used while nursing;      (2d) Other specific use restrictions as noted.

Class: 2b [30]; 2d - Not for long-term use except under the supervision of a qualified practitioner [30].
Ed. Note: Large doses are toxic and can cause vomiting and diarrhea [10, 18, 19].

*Barosma betulina* (Thunb.) Bartl. & Wendl.   Rutaceae
Synonym: *Agathosma betulina* (Bergius) Pillans
*Barosma crenulata* (L.) Hook.
Synonym: *Agathosma crenulata* (L.) Pillans
*Barosma serratifolia* (Curtis) Willd.
Common Name: buchu
Part: leaf
Class: 2b [5]; 2d - Contraindicated in kidney inflammation [30].

*Bauhinia forficata* Link   Fabaceae
Common Name: *pata de vaca*
Part: leaves
Class: 4
Ed. Note: This particular species is not found in any of our primary references. Tannins, flavonoid glycosides, catechols, chalcones, and alkaloids are reported in some of the more than 500 species of the genus (Schultes and Raffauf, 1990). Arvigo and Balick list B. *herrerae*, with common names including *pata de vaca*, as "...an old remedy for birth control..." and declare that drinking a decoction of the plant "...during 9 menstrual cycles is said to produce irreversible infertility in women" (Arvigo and Balick, 1993).

*Benincasa hispida* (Thunb.) Cogn.   Cucurbitaceae
Common Name: wax gourd, *dong gua pi*
Part: rind
Class: 1

*Berberis vulgaris* L.   Berberidaceae
Common Name: barberry
Other Common Names: European barberry
Part: root, root bark
Class: 2b [30, (Mitchell et al., 1983)]
Notice: Berberine (root bark: 6.1%; woody portion of root: 0.4%) [8, 17, 18]
*See page 136.*

---

Class 3: Herbs for which significant data exist to recommend the following labeling: "To be used only under the supervision of an expert qualified in the appropriate use of this substance." Labeling must include proper use information: dosage, contraindications, potential adverse effects and drug interactions, and any other relevant information related to the safe use of this substance.
Class 4: Herbs for which insufficient data are available for classification.

**Ed. Note:** Commission E [4] asserts that berberine is well tolerated up to 0.5 grams, equivalent to over 8.0 grams of the root bark at the 6.1% alkaloid level reported by De Smet [8].

Van Hellemont [24] states that high doses are dangerous, with no quantifying data.

Canadian regulations do not allow barberry in food, except to impart flavor to alcoholic beverages if the material is berberine-free [3].

### *Betula lenta* L. — Betulaceae
**Common Name:** sweet birch
**Other Common Names:** black birch, cherry birch
**Part:** leaf, bark
**Class:** 1
**Notice:** Salicylates (bark: 0.6%) [10, 17] *See page 154.*

### *Betula pendula* Roth. — Betulaceae
### *Betula pubescens* Ehrh.
**Common Name:** white birch
**Part:** leaf, bark
**Class:** 1
**Notice:** Salicylates [26] *See page 154.*
**Ed. Note:** Although Commission E [4] contraindicates *Betula* leaves in irrigation therapy when edema is caused by impaired kidney or heart function, this concern is not relevant to oral consumption; this reference does not list the bark.

### *Bixa orellana* L. — Bixaceae
**Common Name:** annatto
**Part:** seed
**Class:** 1

### *Borago officinalis* L. — Boraginaceae
**Common Name:** borage
**Part:** herb
**Class:** 2a [8, 28, 30]; 2b [8, 9]; 2c [8, 9]; 2d -long-term use is not recommended [30].
**Notice:** Pyrrolizidine Alkaloids (2.0-10.0 ppm) [4, 8, 9, 17, 28, (Awang, 1990)] *See page 149.*

---

Class 1: Herbs that can be safely consumed when used appropriately.
Class 2: Herbs for which the following use restrictions apply, unless otherwise directed by an expert qualified in the use of the described substance:
  (2a) For external use only;   (2b) Not to be used during pregnancy;
  (2c) Not to be used while nursing;   (2d) Other specific use restrictions as noted.

Ed. Note: Effective July 1996, the AHPA Board of Trustees recommends that all products with botanical ingredient(s) which contain toxic pyrrolizidine alkaloids, including *Borago officinalis*, display the following cautionary statement on the label:

"For external use only. Do not apply to broken or abraded skin. Do not use when nursing."

The major alkaloids in *Borago* are reported to be lycopsamine, amabaline, supinine, and 7-acetyllycopsamine [9, (Roitman, 1983), (Larson et al., 1984)]. At least two of these are reported by De Smet [8, 9] to be classified as toxic. However, within the unsaturated PAs, toxicity ranges from mild to severe. Lycopsamine is a mono-esterified, unsaturated PA which is less toxic than di-esterified, unsaturated PAs which occur in other PA-containing plants.

**Part:** seed oil
**Class:** 1

Ed. Note: Borage oil is commonly traded as a source of gamma-linolenic acid (GLA). Pyrrolizidine alkaloids are not detected in the seed oil at a detection limit of 5.0 ppm and are reportedly absent even at 0.5 ppb [9]. The seeds themselves contain only small amounts of the saturated (non-toxic) pyrrolizidine alkaloid, thesinine [9, (Awang, 1990)].

*Boswellia carteri* **Birdwood**                               Burseraceae
*Boswellia serrata* **Roxb. ex Colebr.**
**Common Name:** frankincense
**Other Common Names:** olibanum
**Part:** gum resin
**Class:** 1


*Brassica juncea* **(L.) Czern.**                               Brassicaceae
*Brassica nigra* **(L.) W.D.J. Koch**
**Common Name:** black mustard (*B. nigra*)
**Other Common Names:** mustard
**Part:** seed
**Class Internal Use:** 1

Ed. Note: The classifications and concerns for this herb are based upon therapeutic use and may not be relevant to its consumption as a spice.

Ingestion of a large quantity can cause irritant poisoning [25].

**Class External Use: 2d** - Duration of use should not exceed two weeks [4]; not for external use with children under 6 years of age [10, 18, 25, 28].

Ed. Note: Severe burns occur if applied for a prolonged period (over 15-30

---

Class 3: Herbs for which significant data exist to recommend the following labeling: "To be used only under the supervision of an expert qualified in the appropriate use of this substance." Labeling must include proper use information: dosage, contraindications, potential adverse effects and drug interactions, and any other relevant information related to the safe use of this substance.
Class 4: Herbs for which insufficient data are available for classification.

minutes of the pure mustard powder; when cut with cornstarch or other carrier, potency is reduced).

### *Bupleurum chinense* DC.            Apiaceae
### *Bupleurum falcatum* L.
### *Bupleurum scorzoneraefolium* Willd.
**Common Name:** Chinese thoroughwax (*B. chinense*), *chai hu*
**Part:** root
**Class:** 1

### *Buxus sempervirens* L.            Buxaceae
**Common Name:** boxwood
**Part:** leaf
**Class:** 3 [12, 15, 18, 19, 24]
**Ed. Note:** Although *Buxus* leaf is reported as an adulterant of uva-ursi [18], Wichtl [27] states that this is "...rarely found to occur in practice."

### *Calendula officinalis* L.            Asteraceae
**Common Name:** calendula, pot marigold
**Part:** flower
**Class:** 1

### *Calluna vulgaris* (L.) Hull            Ericaceae
**Common Name:** heather
**Part:** flower
**Class:** 1

### *Camellia sinensis* (L.) Kuntze            Theaceae
**Common Name:** tea
**Other Common Names:** black tea, green tea, Chinese tea
**Part:** leaf, stems
**Class:** 2d - Fermented black teas are not recommended for excessive or long-term use [30].
**Ed. Note:** Green, unfermented tea is the world's second most popular beverage, after water (Graham, 1992).
**Notice:** Nervous System Stimulant (0.9-5.0% caffeine) [6, 13, 17, 18, 21, 27, 28] *See page 174.*
Tannins (green tea: 22.2%, black tea: 12.9%) [13, 18, 21, 27] *See page 155.*

---

Class 1: Herbs that can be safely consumed when used appropriately.
Class 2: Herbs for which the following use restrictions apply, unless otherwise directed by an expert qualified in the use of the described substance:
    (2a) For external use only;     (2b) Not to be used during pregnancy;
    (2c) Not to be used while nursing;     (2d) Other specific use restrictions as noted.

***Cananga odorata* (Lam.)** J. D. Hook. & T. Thompson.     **Annonaceae**
**Common Name:** ylang ylang
**Part:** seed
**Class:** 4


***Canarium album* (Lour.) Raeush.**     **Burseraceae**
**Common Name:** white Chinese olive, *qing guo*
**Part:** fruit
**Class:** 1


***Capsella bursa-pastoris* (L.) Medik.**     **Brassicaceae**
**Common Name:** shepherd's purse
**Part:** herb
**Class:** 2b [6, 25]; 2d - Individuals with a history of kidney stones should use cautiously [30].
**Notice:** Oxalates [13] *See page 148.*
Emmenagogue/Uterine Stimulant [4, 28] *See page 169.*
**Ed. Note:** Large doses of extract may cause heart palpitations [30].


***Capsicum annuum* L. var. *annuum***     **Solanaceae**
***Capsicum annuum* L. var. *glabriusculum* (Dunal) Heiser & Pickersgill**
***Capsicum annuum* var. *frutescens* (L.) Kuntze**
**Common Name:** cayenne
**Part:** fruit
**Class, Internal Use:** 1
**Ed. Note:** Excessive doses may cause GI irritation in sensitive individuals [6, 10, 21, 24, 25].

**Class, External Use:** 2d - Contraindicated on injured skin or near eyes [4, 17, 24, 25, 30].
**Ed. Note:** Commission E [4] suggests that *Capsicum* should not be applied externally for more than 2 days, with a 14-day time lapse between applications. This reference also notes that continued use on the same area may cause damage to sensitive nerves. This information is contradicted by the *Physicians Desk Reference* (Huff, 1989) listing for Zostrex, which states that capsaicin-containing preparations must be used continuously for up to several weeks to be effective.

---

Class 3: Herbs for which significant data exist to recommend the following labeling: "To be used only under the supervision of an expert qualified in the appropriate use of this substance." Labeling must include proper use information: dosage, contraindications, potential adverse effects and drug interactions, and any other relevant information related to the safe use of this substance.
Class 4: Herbs for which insufficient data are available for classification.

***Carica papaya* L.**     Caricaceae
Common Name: papaya
Part: leaf
Class: 1

***Carthamus tinctorius* L.**     Asteraceae
Common Name: safflower, *hong hua*
Part: flower
Class: 2b [2, 6, 7, 18, 23, 29]; 2d - Contraindicated in patients with hemorrhagic diseases or peptic ulcers [2, 7, 14].
Notice: Abortifacient [18] *See page 163.*
Emmenagogue/Uterine Stimulant [2, 13, 14] *See page 169.*
Ed. Note: May prolong blood coagulation time [14].

***Carum carvi* L.**     Apiaceae
Common Name: caraway
Part: fruit (commonly know as "seed")
Class: 1

***Castanea dentata* (Marsh.) Borkh.**     Fagaceae
Common Name: American chestnut
Part: leaf
Class: 4

***Castanea sativa* Mill.**     Fagaceae
Common Name: Spanish chestnut
Part: leaf
Class: 1
Notice: Tannins (9.0%) [18, 27, 30] *See page 155.*

***Catharanthus roseus* (L.) G. Don.**     Apocynaceae
Common Name: Madagascar periwinkle
Part: herb
Class: 3 [14, 18, 21]
Notice: Abortifacient [18] *See page 163.*

***Caulophyllum thalictroides* Michx.**     Berberidaceae
Common Name: blue cohosh
Part: root

---

Class 1: Herbs that can be safely consumed when used appropriately.
Class 2: Herbs for which the following use restrictions apply, unless otherwise directed by an expert qualified in the use of the described substance:
   (2a) For external use only;     (2b) Not to be used during pregnancy;
   (2c) Not to be used while nursing;     (2d) Other specific use restrictions as noted.

**Class:** 2b [3, 16, 30]
**Notice:** Abortifacient [9] *See page 163.*
Emmenagogue/Uterine Stimulant [16, 17, 28]. *See page 169.*
**Ed. Note:** De Smet classifies *Caulophyllum* as an abortifacient. However, eclectic physicians of the early 20th century and practicing midwives of today recommend small amounts of *Caulophyllum* during pregnancy for threatened abortion, and although contraindicated in pregnancy above, *Caulophyllum* may be used as a parturifacient near term to induce childbirth under the supervision of a qualified practitioner [10].

Pilcher's observations (published in JAMA 67:490 in 1916) noted that although the herb has "...a marked stimulating effect on the uterine muscle [of excised uterus of guinea pig] [he] did not believe this would occur under clinical use" (Osol & Farrar, 1955).

Canadian regulations do not allow this herb as a non-medicinal ingredient for oral use products (Michols, 1995) and require that products containing *Caulophyllum* be labeled as not for use in pregnancy.

## *Ceanothus americanus* L.   Rhamnaceae
**Common Name:** New Jersey tea
**Part:** root, root bark
**Class:** 1

## *Centaurea cyanus* L.   Asteraceae
**Common Name:** cornflower
**Part:** herb
**Class:** 1
**Ed. Note:** The *Wealth of India* [6] mistakenly cites Wren's 1956 edition as stating the flowers of *Centaurea* act as an emmenagogue.

## *Centaurium erythraea* Raf.   Gentianaceae
**Synonym:** *Centaurium umbellatum*
**Common Name:** centaury
**Part:** flowering tops
**Class:** 1
**Ed. Note:** Although regulated in the U.S. as an allowable flavoring agent in alcoholic beverages only [20], no other reference records any such limitation on use.

---

Class 3: Herbs for which significant data exist to recommend the following labeling: "To be used only under the supervision of an expert qualified in the appropriate use of this substance." Labeling must include proper use information: dosage, contraindications, potential adverse effects and drug interactions, and any other relevant information related to the safe use of this substance.
Class 4: Herbs for which insufficient data are available for classification.

### *Centella asiatica* (L.) Urban.     Apiaceae
**Common Name:** gotu kola
**Part:** herb
**Class:** 1
**Ed. Note:** Canadian regulations do not allow gotu kola as a non-medicinal ingredient for oral use products (Michols, 1995).

### *Cephaelis ipecacuanha* (Brot.) Tussac.     Rubiaceae
**Common Name:** ipecac
**Part:** rhizome
**Class:** 2b [30]; 2d - May cause nausea and vomiting [17, 20, 21, 27]; not for long-term use [5, 17, 21, 24, 26, 27]; contraindicated in cardiac disease (Mitchell et al., 1983; Gilman et al., 1985).
**Standard Dose:** The therapeutic dose of ipecac as an expectorant is achieved at an adult dose of 0.4-1.4 ml of Ipecac Syrup, USP [21]. The emetic dose of Ipecac Syrup, USP is 15.0 ml (1 tablespoon) [20, 21, 27].
**Notice:** Emetic [17, 20, 21, 27] *See page 167.*
**Ed. Note:** Ipecac Syrup is allowed to be sold in the U.S. only in packages of less than 1.0 ounce and must be labeled with warnings against use in children under the age of 1 year, in unconscious persons, or following the consumption of "...strychnine, corrosives such as alkalies (lye) and strong acids, or petroleum distillates such as...paint thinner." Packaging must also direct the user to call a physician or hospital prior to use [20].

Concentrated preparations such as fluid extracts of ipecac can be extremely toxic [21, (Gilman et al., 1985)] and should be used only under proper supervision [20, 30].

### *Ceratonia siliqua* L.     Fabaceae
**Common Name:** carob
**Part:** fruit
**Class:** 1

### *Cetraria islandica* (L.) Achar.     Parmeliaceae
**Common Name:** Iceland moss
**Part:** thallus
**Class, Decoction or Infusion (teas):** 1
**Class, Alcohol Extract or Powder:** 2d - Contraindicated in gastro-duodenal ulcers due to mucosa-irritating property [24, 30].

---

Class 1: Herbs that can be safely consumed when used appropriately.
Class 2: Herbs for which the following use restrictions apply, unless otherwise directed by an expert qualified in the use of the described substance:
   (2a) For external use only;     (2b) Not to be used during pregnancy;
   (2c) Not to be used while nursing;     (2d) Other specific use restrictions as noted.

**Notice:** GI Irritant [24, 30] *See page 172.*
**Ed. Note:** Regulated in the U.S. as an allowable flavoring agent in alcoholic beverages only [20].

## *Chaenomeles speciosa* (Sweet) Nakai.  Rosaceae
**Common Name:** flowering quince, *mu gua*
**Part:** fruit
**Class:** 1
**Ed. Note:** Bensky & Gamble [2] state that excessive use can damage teeth and bones.

## *Chamaelirium luteum* (L.) A. Gray.  Liliaceae
**Common Name:** false unicorn root
**Other Common Names:** helonias
**Part:** rhizome
**Class:** 2b [5]
**Notice:** Emmenagogue/Uterine Stimulant [5] *See page 169.*
GI Irritant [5, 18] *See page 172.*
**Ed. Note:** Canadian regulations do not allow false unicorn root as a non-medicinal ingredient for oral use products (Michols, 1995).

## *Chamaemelum nobile* (L.) All.  Asteraceae
**Common Name:** Roman chamomile
**Part:** flower
**Class:** 2b [18]
**Notice:** Abortifacient [18] *See page 163.*
Emmenagogue/Uterine Stimulant [18] *See page 169.*
**Ed. Note:** Although Leung & Foster [17] cite the 1982 edition of *Martindale: The Extra Pharmacopoeia* as the source for a concern for emesis associated with the consumption of Roman chamomile, no such concern is stated in the later edition of *Martindale: The Extra Pharmacopoeia* [21].

## *Chamomilla suaveolens* (Pursh) Rydb.  Asteraceae
**Common Name:** pineapple weed
**Part:** flowering tops
**Class:** 1

---

Class 3: Herbs for which significant data exist to recommend the following labeling: "To be used only under the supervision of an expert qualified in the appropriate use of this substance." Labeling must include proper use information: dosage, contraindications, potential adverse effects and drug interactions, and any other relevant information related to the safe use of this substance.
Class 4: Herbs for which insufficient data are available for classification.

### *Changium smyrnoides* Wolff   Apiaceae
Common Name: *ming dang shen*
Part: root
Class: 2b [29]

### *Chelidonium majus* L.   Papaveraceae
Common Name: celandine, *bai qu cai*
Other Common Names: greater celandine
Part: herb
Class: 2b [30]; 2d - Not to be used by children [24].
Standard Dose: 600.0 mg to 5.0 grams per day [4, 26, 27].
Notice: Berberine [28, (Duke, 1992), (Osol & Farrar, 1955)] *See page 136.*
Ed. Note: Although severe irritation of the digestive system is recorded in association with consumption of the *fresh* plant [15, 16, 18], the herb in its dried form is approved for internal use in Germany with no contraindications or side effects noted [4]. "In animal experiments, however, no such symptoms and no pathological changes in the digestive organs have been seen. More than 500 grams of greater celandine is said to be required to cause toxic effects in horses and cattle" [11].

Muenscher's [19] report of toxicity, attributed to Milspaugh (1887), is based on homeopathic provings in which the herb is administered in large doses specifically intended to induce symptoms. This is not relevant to regular usage patterns.

Canadian regulations do not allow celandine in foods [3]. An admonition against exceeding the stated dosage has been proposed in Australia, and the following labeling has been suggested: "May affect glaucoma treatment" [1].

### *Chelone glabra* L.   Scrophulariaceae
Common Name: balmony
Part: flowering tops
Class: 1

### *Chimaphila umbellata* (L.) W. Barton   Pyrolaceae
Common Name: pipsissewa
Other Common Names: prince's pine
Part: herb
Class: 1

---

Class 1: Herbs that can be safely consumed when used appropriately.
Class 2: Herbs for which the following use restrictions apply, unless otherwise directed by an expert qualified in the use of the described substance:
    (2a) For external use only;    (2b) Not to be used during pregnancy;
    (2c) Not to be used while nursing;    (2d) Other specific use restrictions as noted.

*Chionanthus virginicus* L.  Oleaceae
Common Name: fringetree
Part: root
Class: 1


*Cibotium barometz* (L.) J. Sm.  Thyrsopteridaceae
Common Name: scythian lamb, *gou ji*
Part: rhizome
Class: 1


*Cichorium intybus* L.  Asteraceae
Common Name: chicory
Part: root
Class: 1
Ed. Note: Although Commission E [4] suggests that a physician be consulted prior to use by those with a history of gallstones, no other literature supports this concern.
   Allergic skin reactions are cited "in rare cases", most likely due to the presence of irritating latex associated with handling of fresh plant material [4, 30].


*Cimicifuga foetida* L.  Ranunculaceae
Common Name: bugbane, *sheng ma*
Part: rhizome
Class: 2d - Contraindicated in measles and in persons with breathing difficulties [2]; do not exceed recommended dose [30].
Standard Dose: 1.5-9.0 grams daily, as a tea [2, 13, (Yen, 1992)].
Ed. Note: Symptoms of overdose may include "...headaches, dizziness, vomiting, tremors, gastroenteritis, and pathogenic erections" [2]. The contained constituent cimicifugine can produce gastroenteritis [29].


*Cimicifuga racemosa* (L.) Nutt.  Ranunculaceae
Common Name: black cohosh
Other Common Names: black snakeroot
Part: rhizome
Class: 2b [3, 5, 18]; 2c [5, 18]
Notice: Emmenagogue/Uterine Stimulant [30] *See page 169*.
Ed. Note: Occasional gastrointestinal discomfort may occur [5], and large

---

Class 3: Herbs for which significant data exist to recommend the following labeling: "To be used only under the supervision of an expert qualified in the appropriate use of this substance." Labeling must include proper use information: dosage, contraindications, potential adverse effects and drug interactions, and any other relevant information related to the safe use of this substance.
Class 4: Herbs for which insufficient data are available for classification.

doses may cause vertigo, headache, nausea, impaired vision, vomiting, and impaired circulation [10, 18]. An estrogenic effect and lowering of blood pressure have been recorded [18].

Commission E [4] recommends that use be limited to six months.

Canadian regulations do not allow black cohosh as a non-medicinal ingredient for oral use products (Michols, 1995).

*Cinchona calisaya* Wedd.   Rubiaceae
*Cinchona ledgeriana* Moens & Trimen.
*Cinchona officinalis* L.
*Cinchona pubescens* Vahl.

**Common Name:** quinine, yellow quinine (*C. ledgeriana*), red cinchona (*C. pubescens*)

**Other Common Names:** calisaya bark (*C. calisaya*), yellow cinchona (*C. ledgeriana*)

**Part:** bark

**Class:** 2b [4, 5, 18, 27]; 2d - Contraindicated in gastrointestinal ulcers [27]; do not exceed recommended dose [30].

**Standard Dose:** Internally: 0.6-1 gram per day [Merck & Co., 1930].

**Ed. Note:** Rare adverse events of bleeding associated with platelet reduction are reported, with a warning that simultaneous consumption may potentiate the anticoagulant action of coumarin derivatives [4, 9].

Symptoms of "cinchonism", reportedly due to overdose of the alkaloid quinine, include headache, nausea, disturbed vision, tinnitus, delirium, abdominal pain, and diarrhea. Extreme reactions including blindness, deafness, convulsions, paralysis, collapse, and even death may occur in hypersensitive individuals [4, 5, 17, 20, 21, 24].

Regulated in the U.S. as allowable in carbonated beverages in certain salt forms as long as content of quinine does not exceed 83.0 ppm and as long as the ingredient is identified on the label as "quinine". U.S. drug regulations require the following label on all *Cinchona* derivatives: "Discontinue use if ringing in the ears, deafness, skin rash or visual disturbances occur" [20].

*Cinnamomum camphora* (L.) J. Presl.   Lauraceae
**Common Name:** camphor
**Part:** distillate of the wood
**Class:** 2b [25]; 2d - Camphor preparations should not be used in the facial regions of infants and small children, especially not in the nasal area [4, 11, 21]; not for long-term use; do not exceed recommended dose [1, 30].

---

Class 1: Herbs that can be safely consumed when used appropriately.
Class 2: Herbs for which the following use restrictions apply, unless otherwise directed by an expert qualified in the use of the described substance:
    (2a) For external use only;    (2b) Not to be used during pregnancy;
    (2c) Not to be used while nursing;    (2d) Other specific use restrictions as noted.

**Standard Dose:** Externally: In concentration of less than 10.0% as Spirit of Camphor, less than 20.0% in semi-solid preparations. Internally: 30 to 300 mg per day [4].
**Notice:** Safrole [6, 13] *See page 152.*
**Ed. Note:** Dosages exceeding 2.0 grams have been reported to be narcotic and may cause convulsions, delirium, hallucinations, and death [6, 10, 14, 18, 25].

Generally recognized as safe in the U.S. if safrole-free [20].

## *Cinnamomum cassia* J. Presl. — Lauraceae
**Synonym:** *Cinnamomum aromaticum*
**Common Name:** cassia, *rou gui*
**Other Common Names:** cassia lignea, cassia cinnamon
**Part:** bark
**Class:** 2b [2, 4, 10, 18]
**Standard Dose:** 1.0-4.5 grams daily, as a tea [2, 4, 13].
**Ed. Note:** The classifications and concerns for this herb are based upon therapeutic use and may not be relevant to its consumption as a spice.

Contrary to the 2b classification above, De Smet [8] states that "cinnamon does not present any special risk in pregnancy" and that abortifacient reports in Roth [22] and List & Hörhammer [18] do not contain concrete case citations. He concludes, however, that "prolonged use of the essential oil should be restricted during pregnancy."

Allergic reactions of skin and mucosa have been reported, of special concern for persons prone to allergic dermatitis and in products such as ointments, toothpaste, and mouthwash where the content of cinnamaldehyde is above 0.01% [4, 8].

There is some suggestion that tetracycline and related substances should not be administered simultaneously to therapeutic dosage levels of cassia, due to suspected retardation of the drugs' absorption [8].

**Common Name:** *gui zhi*
**Part:** twigs
**Class:** 1

## *Cinnamomum verum* J. Presl. — Lauraceae
**Synonym:** *Cinnamomum zeylanicum* Blume
**Common Name:** cinnamon
**Other Common Names:** Ceylon cinnamon, Seychelles cinnamon
**Part:** bark

---

Class 3: Herbs for which significant data exist to recommend the following labeling: "To be used only under the supervision of an expert qualified in the appropriate use of this substance." Labeling must include proper use information: dosage, contraindications, potential adverse effects and drug interactions, and any other relevant information related to the safe use of this substance.
Class 4: Herbs for which insufficient data are available for classification.

**Class:** 2b [6, 22, 4]; 2d - Not for long-term use; do not exceed recommended dose [30].
**Standard Dose:** 2.0 to 4.0 grams of bark daily, as tea; 0.05 to 0.2 grams daily of essential oil [4].
**Notice:** Safrole (up to 0.25%) [8, 17, (Heath, 1981)] *See page 152.*
**Ed. Note:** The classifications and concerns for this herb are based upon therapeutic use and may not be relevant to its consumption as a spice.

The notes attributed to De Smet [8] in the listing for *C. cassia* regarding its use in pregnancy are also relevant to this species.

Large doses have been reported to cause side effects such as stimulation of the vasomotor center [27] and methemoglobinemia and hematinemia with subsequent nephritis [22]. This last reference, however, is brought into question by De Smet [8], who notes that the original report was of a single incident of consumption of an unspecified "large amount" of a potentially misidentified substance [8].

Several species of *Cinnamomum* are interchangeable in trade as cinnamon [17].

### *Cistanche salsa* (C. A. Mey.) Beck.          Orobanchaceae
**Common Name:** broomrape, *rou cong rong*
**Part:** root
**Class:** 1

### *Citrus aurantifolia* (Christm.) Swingle.          Rutaceae
**Common Name:** lime
**Part:** peel
**Class:** 1

### *Citrus aurantium* L.          Rutaceae
**Common Name:** bitter orange
**Other Common Names:** sour orange, Seville orange
**Part:** peel
**Class:** 1
**Notice:** Photosensitizing [4] *See page 176.*
**Ed. Note:** Bensky & Gamble [2] cite caution during pregnancy, a concern not duplicated in other references or for any other *Citrus* species.

Significant adverse responses ("intestinal colic, convulsions, and even death") are reported in children following ingestion of large amounts of orange peel [17].

---

Class 1: Herbs that can be safely consumed when used appropriately.
Class 2: Herbs for which the following use restrictions apply, unless otherwise directed by an expert qualified in the use of the described substance:
    (2a) For external use only;      (2b) Not to be used during pregnancy;
    (2c) Not to be used while nursing;      (2d) Other specific use restrictions as noted.

*Citrus bergamia* Risso & Poit.    Rutaceae
Common Name: bergamot orange
Part: peel
Class: 1

*Citrus limon* (L.) N. L. Burm.    Rutaceae
Common Name: lemon
Part: peel
Class: 1

*Citrus limonia* Osbeck.    Rutaceae
Common Name: mandarin lime
Other Common Names: lemandarin
Part: peel
Class: 1

*Citrus reticulata* Blanco.    Rutaceae
Common Name: tangerine, *qing pi* (immature peel), *chen pi* (aged mature peel)
Other Common Names: mandarin orange
Part: peel
Class: 1

*Clematis chinensis* Osbeck.    Ranunculaceae
Common Name: Chinese clematis, *wei ling xian*
Part: root
Class: 1
Ed. Note: The sap is reported to contain protoanemonin, a vessicant oil, which is a direct irritant of the skin and mucous membranes [7, 15, 16, 18]. The irritant principles are mostly lost upon drying [18].

*Cnicus benedictus* L.    Asteraceae
Common Name: blessed thistle
Other Common Names: holy thistle
Part: herb
Class: 2b [5, 10]
Ed. Note: High doses (more than 5.0 grams per cup of tea [24]) may irritate the stomach and cause vomiting [22, 24].

Regulated in the U.S. as an allowable flavoring agent in alcoholic beverages only [20].

---

Class 3: Herbs for which significant data exist to recommend the following labeling: "To be used only under the supervision of an expert qualified in the appropriate use of this substance." Labeling must include proper use information: dosage, contraindications, potential adverse effects and drug interactions, and any other relevant information related to the safe use of this substance.
Class 4: Herbs for which insufficient data are available for classification.

***Cnidium monnieri*** **(L.) Cuss.**      **Apiaceae**
Common Name: cnidium, *she chuang zi*
Part: seed
Class: 1


***Codonopsis pilosula*** **(Franch.) Nannfeldt.**      **Campanulaceae**
***Codonopsis tangshen*** **Oliver.**
Common Name: codonopsis, *dang shen*
Other Common Names: bellflower
Part: root
Class: 1


***Coffea arabica*** **L.**      **Rubiaceae**
Common Name: coffee
Part: roasted seed kernel
Class: 2b [(Anonymous, *Consumer Reports*, 1987), (Infante-Rivard, 1993)];
2d - Not recommended for excessive or long-term use [30]; contraindicated for people with gastric ulcer [17]; contraindicated in glaucoma (temporarily increases intraocular pressure) [17].
Notice: Nervous System Stimulant (usually 1.5-2.5% caffeine) [17] *See page 174.*
GI Irritant [6, 10, 17, 21] *See page 172.*
Ed. Note: Side effects cited in the literature, even for moderate use, remain controversial. These include chronic headaches, hypertension, diuresis, possible increased risk of pancreatic, breast (in obese women), and ovarian cancer [17], increased cholesterol levels (Anonymous, 1987; Welland & Gray, 1996), increased incidence of delayed conception (Stanton, 1995), increased risk of fetal loss (Infante-Rivard, 1993), increased excretion of calcium and magnesium and potential for osteoporosis (Stavric, 1992; Welland & Gray, 1996), and increased risk of heart attack in women (Leonard et al., 1987).


***Coix lacryma-jobi*** **L.**      **Poaceae**
Common Name: Job's tears, *yi yi ren*
Part: seed
Class: 2b [2, 14, 29]


***Cola acuminata*** **(P. Beauv.) Schott. & Endl.**      **Sterculiaceae**
***Cola nitida*** **Schott. & Endl.**
Common Name: cola
Other Common Names: kola nut, bissy nut

---

Class 1: Herbs that can be safely consumed when used appropriately.
Class 2: Herbs for which the following use restrictions apply, unless otherwise directed by an expert qualified in the use of the described substance:
    (2a) For external use only;      (2b) Not to be used during pregnancy;
    (2c) Not to be used while nursing;      (2d) Other specific use restrictions as noted.

**Part:** seed
**Class:** 2b [30]; **2d** - Contraindicated in hypertension [5] and gastric and duodenal ulcers [4]; not recommended for excessive or long-term use [30].
**Notice:** Nervous System Stimulant (1.0-2.5% caffeine) [6, 17, 18, 21] *See page 174.*
GI Irritant [4] *See page 172.*

### *Collinsonia canadensis* L. — Lamiaceae
**Common Name:** collinsonia
**Other Common Names:** richweed, stoneroot
**Part:** root
**Class:** 1

### *Commiphora madagascariensis* Jacq. — Burseraceae
### *Commiphora molmol* Engl. ex Tschirch
### *Commiphora myrrha* (Nees) Engl.
**Common Name:** myrrh
**Part:** gum resin
**Class:** 2b [2, 17]; **2d** - Contraindicated in excessive uterine bleeding [2].
**Notice:** Emmenagogue/Uterine Stimulant [17] *See page 169.*
**Ed. Note:** Doses over 2.0-4.0 grams may cause irritation of the kidneys and diarrhea [18, 24].
Permitted for external use only in France [9].

### *Commiphora mukul* (Hook. ex Stocks) Engl. — Burseraceae
**Common Name:** guggul
**Part:** prepared gum resin
**Class:** 2b [6, 17]
**Notice:** Emmenagogue/Uterine Stimulant [6, 17] *See page 169.*

### *Convallaria majalis* L. — Liliaceae
**Common Name:** lily of the valley
**Part:** entire plant
**Class:** 3 [1, 4, 11, 12, 16, 18, 21, 22, 24]
**Notice:** Cardiac Glycosides [1, 12] *See page 138.*
**Ed. Note:** Canadian regulations do not allow lily of the valley in foods [3].

---

Class 3: Herbs for which significant data exist to recommend the following labeling: "To be used only under the supervision of an expert qualified in the appropriate use of this substance." Labeling must include proper use information: dosage, contraindications, potential adverse effects and drug interactions, and any other relevant information related to the safe use of this substance.
Class 4: Herbs for which insufficient data are available for classification.

*Conyza canadensis* (L.) Cronq.　　　　　　　　　　Asteraceae
**Common Name:** Canada fleabane
**Part:** herb
**Class:** 1


*Coptis chinensis* Franch.　　　　　　　　　　Ranunculaceae
**Common Name:** Chinese goldthread, *huang lian*
**Part:** rhizome
**Class:** 2b [30]
**Notice:** Berberine (4.0-7.0%) [2, 7, 13, 29] *See page 136.*
Emmenagogue/Uterine Stimulant [2, 7, 13, 29] *See page 169.*


*Coptis groenlandica* (Oed.) Fern.　　　　　　　　　　Ranunculaceae
**Common Name:** goldthread
**Other Common Names:** canker root
**Part:** rhizome
**Class:** 2b [30]
**Notice:** Berberine [28, (Solis-Cohen & Githens, 1928)] *See page 136.*


*Cordia salicifolia* Cham.　　　　　　　　　　Boraginaceae
**Common Name:** cha-de-bugre
**Part:** leaf
**Class:** 4


*Cordyceps sinensis* (Berk.) Sacc.　　　　　　　　　　Ascomycetes
**Common Name:** cordyceps, *dong chong zia cao*
**Part:** fruiting body and larvae
**Class:** 1


*Coriandrum sativum* L.　　　　　　　　　　Apiaceae
**Common Name:** coriander (fruit), cilantro (leaf)
**Other Common Names:** Chinese parsley
**Part:** fruit (commonly know as "seed") and leaf
**Class:** 1

---

Class 1: Herbs that can be safely consumed when used appropriately.
Class 2: Herbs for which the following use restrictions apply, unless otherwise directed by an expert qualified in the use of the described substance:
　　(2a) For external use only;　　(2b) Not to be used during pregnancy;
　　(2c) Not to be used while nursing;　　(2d) Other specific use restrictions as noted.

*Cornus officinalis* Sieb. & Zucc.     Cornaceae
**Common Name:** Asiatic dogwood, *shan zhu yu*
**Other Common Names:** Japanese cornel, dogwood fruit
**Part:** fruit
**Class:** 2d - Contraindicated in individuals with painful or difficult urination [2].


*Corydalis yanhusuo* W.T. Wang.     Fumariaceae
**Common Name:** corydalis, *yan hu suo*
**Part:** rhizome
**Class:** 2b [2, 18, 29]
**Notice:** Emmenagogue/Uterine Stimulant [18] *See page 169.*


*Corylus avellana* L.     Betulaceae
*Corylus cornuta* Marsh.
**Common Name:** European hazel
**Other Common Names:** European filbert (*C. avellana*), beaked hazel (*C. cornuta*), beaked filbert (*C. cornuta*), filbert leaf
**Part:** leaf and bark
**Class:** 1


*Crataegus laevigata* (Poir.) DC.     Rosaceae
*Crataegus monogyna* Jacq.
**Common Name:** English hawthorn (*C. laevigata*), oneseed hawthorn (*C. monogyna*)
**Part:** flowers, leaves, fruits
**Class:** 1
**Ed. Note:** Patients prescribed digitalis should be informed that hawthorn preparations may have a potentiating effect which may necessitate a smaller dosage of digitalis [24].


*Crataegus pinnatifida* Bge.     Rosaceae
**Common Name:** Chinese hawthorn, *shan zha*
**Part:** fruit
**Class:** 1
**Ed. Note:** Hsu [13] and Yeung [29] report that *Crataegus* causes uterine contractions but neither contraindicates it in pregnancy.

---

Class 3: Herbs for which significant data exist to recommend the following labeling: "To be used only under the supervision of an expert qualified in the appropriate use of this substance." Labeling must include proper use information: dosage, contraindications, potential adverse effects and drug interactions, and any other relevant information related to the safe use of this substance.
Class 4: Herbs for which insufficient data are available for classification.

### *Crocus sativus* L.  Iridaceae
**Common Name:** saffron
**Other Common Names:** Spanish saffron, true saffron
**Part:** stigmas
**Class:** 2b [6, 11, 18, 23]
**Notice:** Abortifacient [4, 6, 11, 18, 23, 27] *See page 163.*
Emmenagogue/Uterine Stimulant [10, 13, 17] *See page 169.*
**Ed. Note:** Severe side effects occur on ingestion of 5.0 grams of saffron [4, 17, 27]. The lethal dose is given by Wichtl [27] as 20.0 grams.

No risk is associated with consumption in standard food use quantities or in therapeutic dosage of less than 1.5 grams per day [4, 17].

### *Cullen corylifolia* (L.) Medicus.  Fabaceae
**Synonym:** *Psoralea corylifolia* L.
**Common Name:** psoralea, *bu gu zhi*
**Other Common Names:** scurfy pea
**Part:** seed
**Class:** 2b [30]; 2d - Avoid prolonged exposure to sunlight [30].
**Notice:** Abortifacient [14] *See page 163.*
Photosensitizing [7, 14] *See page 176.*
**Ed. Note:** Oral administration of the powdered seeds is reported to generally result in nausea, vomiting, malaise, headache, and sometimes in purging [6].

### *Cuminum cyminum* L.  Apiaceae
**Common Name:** cumin
**Part:** fruit (commonly know as "seed")
**Class:** 1

### *Cuphea balsamona* Cham. & Schlechtend.  Lythraceae
**Common Name:** sete sangrias
**Part:** herb
**Class:** 2b (Penna, 1946); 2d- Not to be used by children (Penna, 1946).

### *Curculigo orchioides* Gaertn.  Amaryllidaceae
**Common Name:** golden eye-grass, *xian mao*
**Part:** rhizome
**Class:** 3 [2]

---

Class 1: Herbs that can be safely consumed when used appropriately.
Class 2: Herbs for which the following use restrictions apply, unless otherwise directed by an expert qualified in the use of the described substance:
    (2a) For external use only;    (2b) Not to be used during pregnancy;
    (2c) Not to be used while nursing;    (2d) Other specific use restrictions as noted.

***Curcuma aromatica* Salisbury.**     Zingiberaceae
***Curcuma domestica* Valet.**
***Curcuma longa* L.**
**Common Name:** turmeric, *jiang huang*
**Other Common Names:** curcuma
**Part:** rhizome
**Class:** 2b [2, 7, 25, 29, 30]; 2d - Therapeutic quantities should not be taken by people with bile duct obstruction or gall stones [4]; *C. longa* should not be administered to patients who suffer from stomach ulcers or hyperacidity [24].
**Standard Dose:** 1.5-3.0 grams daily [4]; 4.5-9.0 grams prepared as tea [2].
**Notice:** Emmenagogue/Uterine Stimulant [7, 25] *See page 169.*
**Ed. Note:** The classifications and concerns for this herb are based upon therapeutic use and may not be relevant to its consumption as a spice.

***Curcuma zedoaria* (Berg.) Roscoe.**     Zingiberaceae
**Common Name:** zedoary, *e zhu*
**Part:** rhizome
**Class:** 2b [2, 29]
**Ed. Note:** Bensky & Gamble [2] advise cautious use during excessive menstruation.

***Cuscuta chinensis* Lam.**     Cuscutaceae
***Cuscuta japonica* Choisy**
**Common Name:** dodder, *tu si zi*
**Other Common Names:** cuscuta
**Part:** seed
**Class:** 1

***Cyamopsis tetragonolobus* (L.) Taubert**     Fabaceae
**Common Name:** guar gum
**Part:** seed
**Class:** 2d - Take with at least 250 ml (8 oz) liquid [17, 20, 21]; contraindicated in bowel obstruction [21, 26].
**Standard Dose:** 5.0 grams up to three times daily, with or preceding meals [21].
**Notice:** Bulk Forming Laxative [26] *See page 165.*
**Ed. Note:** May cause flatulence, diarrhea, or nausea, especially in the early stage of therapeutic use [17, 21].

---

Class 3: Herbs for which significant data exist to recommend the following labeling: "To be used only under the supervision of an expert qualified in the appropriate use of this substance." Labeling must include proper use information: dosage, contraindications, potential adverse effects and drug interactions, and any other relevant information related to the safe use of this substance.
Class 4: Herbs for which insufficient data are available for classification.

A risk of guar gum affecting the absorption of other drugs is reported [21]. Specific labeling is required in the U.S. for all OTC drug products containing guar gum [20] *See Appendix 2, page 165.*

### *Cyathula officinalis* Kuan.            Amaranthaceae
**Common Name:** cyathula, *chuan niu xi*
**Part:** root
**Class:** 2b [2, 29]

### *Cymbopogon citratus* (DC. ex Nees) Stapf.     Poaceae
**Common Name:** lemongrass
**Other Common Names:** West Indian lemongrass, fever grass
**Part:** herb
**Class:** 2b [18]
**Notice:** Emmenagogue/Uterine Stimulant [18] *See page 169.*

### *Cynanchum atratum* Bge.            Asclepidaceae
**Common Name:** cynanchum, *bai wei*
**Part:** root
**Class:** 1

### *Cynomorium songaricum* Rupr.          Cynomoriaceae
**Common Name:** cynomorium, *suo yang*
**Part:** root
**Class:** 2d - Contraindicated in patients with diarrhea [2].

### *Cyperus rotundus* L.            Cyperaceae
**Common Name:** nut grass, *xiang fu*
**Other Common Names:** cyperus
**Part:** rhizome
**Class:** 1
**Ed. Note:** Although *Cyperus* is sometimes classified as an emmenagogue [18, 25], four of the primary references based on traditional Chinese medicine describe a contraction-inhibiting effect on the uterus in pregnant and non-pregnant women [7, 13, 14, 29].

---

Class 1: Herbs that can be safely consumed when used appropriately.
Class 2: Herbs for which the following use restrictions apply, unless otherwise directed by an expert qualified in the use of the described substance:
    (2a) For external use only;     (2b) Not to be used during pregnancy;
    (2c) Not to be used while nursing;     (2d) Other specific use restrictions as noted.

***Cytisus scoparius* (L.) Link.**  Fabaceae
**Common Name:** Scotch broom
**Other Common Names:** broom, broom tops
**Part:** flowering tops
**Class:** 3 [10, 17, 21, 24, 28]
**Notice:** Abortifacient [18] *See page 163.*


***Daemonorops draco* Blume.**  Arecaceae
**Common Name:** dragon's blood, *xue jie*
**Other Common Names:** dragon's blood palm
**Part:** resin
**Class:** 1


***Daucus carota* L.**  Apiaceae
**Common Name:** wild carrot
**Other Common Names:** carrot, Queen Ann's lace
**Part:** fruit (commonly know as "seed")
**Class:** 2b [10, 18, 25]


***Dendranthema* x *morifolium* (Ramat.) Tzelev**  Asteraceae
**Common Name:** mum, *ju hua*
**Other Common Names:** florist's chrysanthemum, chrysanthemum
**Part:** flower
**Class:** 1


***Dendrobium nobile* Lindl.**  Orchidaceae
**Common Name:** dendrobium, *shi hu*
**Part:** herb
**Class:** 1
**Ed. Note:** Large doses can have an inhibitory effect on the heart and lungs, and overdoses can cause convulsions [2]. These actions are most likely due to the presence of the alkaloid dendrobine [23, (Perry, 1980)], found in an amount of 0.3-0.5% in the herb [7, (Yen, 1992)].

---

Class 3: Herbs for which significant data exist to recommend the following labeling: "To be used only under the supervision of an expert qualified in the appropriate use of this substance." Labeling must include proper use information: dosage, contraindications, potential adverse effects and drug interactions, and any other relevant information related to the safe use of this substance.
Class 4: Herbs for which insufficient data are available for classification.

### *Desmodium styracifolium* (Osbeck) Merr.  Fabaceae
**Common Name:** coin-leaved desmodium, *guang jing qian cao*
**Part:** entire plant
**Class:** 4
**Ed. Note:** Listed in only one of our primary references; no contraindications are reported [13].

Substitution with *Lysimachia christinae* and *Glechoma hederacea* is common, and extended consumption of large doses of *Lysimachia* may result in side effects such as dizziness and palpitations due to depletion of potassium associated with the herb's diuretic effect.

### *Digitalis purpurea* L.  Scrophulariaceae
**Common Name:** digitalis
**Other Common Names:** foxglove
**Part:** leaf
**Class:** 3 [1, 6, 8, 10, 15, 16, 18, 21, 25, 26]
**Notice:** Cardiac Glycosides [1, 18, 20] *See page 138.*
**Ed. Note:** U.S. regulations require that digitalis be labeled in a manner which informs consumers of the inappropriateness of use as an antiobesity agent [20]. Canadian regulations do not allow digitalis in foods [3].

### *Dimocarpus longan* Lour.  Sapindaceae
**Synonym:** *Euphoria longan* (Lour.) Steud.
**Common Name:** longan, *long yan rou*
**Part:** fruit
**Class:** 1

### *Dioscorea opposita* Thunb.  Dioscoreaceae
**Common Name:** Chinese yam, *shan yao*
**Other Common Names:** wild yam
**Part:** root
**Class:** 1

### *Dioscorea villosa* L.  Dioscoreaceae
**Common Name:** wild yam, Mexican yam
**Other Common Names:** Atlantic yam
**Part:** rhizome
**Class:** 1
**Ed. Note:** Large doses of the tincture produce emesis [10].

---

Class 1: Herbs that can be safely consumed when used appropriately.
Class 2: Herbs for which the following use restrictions apply, unless otherwise directed by an expert qualified in the use of the described substance:
    (2a) For external use only;    (2b) Not to be used during pregnancy;
    (2c) Not to be used while nursing;    (2d) Other specific use restrictions as noted.

*Dipsacus asper* Wallich.             Dipsacaceae
*Dipsacus japonicus* Miq.
Common Name: teasel, *xu duan*
Other Common Names: Szechuan teasel, Japanese teasel
Part: root
Class: 1

*Dipteryx odorata* (Aubl.) Willd.          Fabaceae
*Dipteryx oppositifolia* Willd.
Common Name: tonka
Other Common Names: tonka bean, Dutch tonka (*D. odorata*), English tonka (*D. oppositifolia*)
Part: seed
Class: 3 [24, 28]
Ed. Note: Allowed in alcoholic beverages in Canada if coumarin-free [3].

*Drynaria fortunei* (Kze.) J.Sm.          Polypodiaceae
Common Name: drynaria, *gu sui bu*
Part: rhizome
Class: 1

*Dryopteris filix-mas* (L.) Schott.          Aspleniaceae
Common Name: male fern
Part: rhizome
Class: 2a [4]; 2b [30]; 2c [30]; 3 [2, 4, 10, 17, 22, 26, 28]
Ed. Note: Canadian regulations do not allow male fern as a non-medicinal ingredient for oral use products (Michols, 1995).

*Echinacea angustifolia* DC.          Asteraceae
Common Name: echinacea angustifolia
Other Common Names: echinacea, narrow-leaved echinacea, Kansas snakeroot, narrow-leaved purple coneflower
Part: root, seed
Class: 1
Ed. Note: Commission E [4] reports for all *Echinacea* species that they are not to be used in systemic diseases such as tuberculosis, leukoses, collagenosis, multiple sclerosis, AIDS, HIV infections, and other autoimmune diseases.
    *Parthenium integrifolium* is recorded as an adulterant of echinacea [30].

---

Class 3: Herbs for which significant data exist to recommend the following labeling: "To be used only under the supervision of an expert qualified in the appropriate use of this substance." Labeling must include proper use information: dosage, contraindications, potential adverse effects and drug interactions, and any other relevant information related to the safe use of this substance.
Class 4: Herbs for which insufficient data are available for classification.

### *Echinacea pallida* (Nutt.) Nutt.             Asteraceae
**Common Name:** echinacea pallida
**Other Common Names:** echinacea, pale-flowered echinacea, pale purple coneflower
**Part:** root, seed
**Class:** 1
**Ed. Note:** Commission E [4] limits duration of use to eight weeks. Also, see *E. angustifolia*.
    *Parthenium integrifolium* is recorded as an adulterant of echinacea [27].

### *Echinacea purpurea* (L.) Moench.          Asteraceae
**Common Name:** echinacea purpurea
**Other Common Names:** echinacea, common echinacea, purple coneflower, Kansas snakeroot
**Part:** root, seed, above-ground parts
**Class:** 1
**Ed. Note:** Commission E [4] recommends a maximum duration of eight weeks for the internal and external use of "...fresh juice and galenic preparations thereof..." of the above-ground parts of this species. Also, see *E. angustifolia*.
    *Parthenium integrifolium* is recorded as an adulterant of echinacea [30].

### *Echinodorus macrophyllus* M. Mich.        Alismataceae
**Common Name:** chapeau de couro
**Part:** leaf, bark, root
**Class:** 1

### *Eclipta prostata* (L.) L.                    Asteraceae
**Common Name:** eclipta, *han lian cao*
**Part:** herb
**Class:** 1

### *Elettaria cardamomum* (L.) Maton.          Zingiberaceae
**Synonym:** *Amomum cardamomum* L.
**Common Name:** cardamom, *bai dou kou*
**Part:** fruit
**Class:** 1

---

Class 1: Herbs that can be safely consumed when used appropriately.
Class 2: Herbs for which the following use restrictions apply, unless otherwise directed by an expert qualified in the use of the described substance:
    (2a) For external use only;     (2b) Not to be used during pregnancy;
    (2c) Not to be used while nursing;     (2d) Other specific use restrictions as noted.

## *Eleutherococcus senticosus* (Rupr. ex Maxim.) Maxim. Araliaceae
**Common Name:** eleuthero, *ci wu jia*
**Other Common Names:** Siberian ginseng, Ussurian thorny pepperbush
**Part:** root, root bark
**Class:** 1
**Ed. Note:** Commission E [4] lists eleuthero as contraindicated in high blood pressure. Farnsworth et al. (1985) quantify this concern, reporting that "In two...studies, it was recommended that the extract not be given to subjects having blood pressure in excess of 180/90 mm Hg." Huang [14], however, states that the glycosides contained in eleuthero lower blood pressure, and that the herb "exerts a tranquilizing effect" on the central nervous system.

Insomnia has been rarely observed in association with clinical studies [9].

While Bradley [5] and Commission E [4] recommend a limitation on the duration of use of from one to three months, the latter states, "...a repeated course is feasible."

*E. senticosus* is commonly adulterated with *Periploca sepium* [21] which contains cardiac glycosides [14] and has been reported to cause hirsuteness in an infant (Koren et al., 1990).

## *Elytrigia repens* (L.) Nevski. Poaceae
**Synonym:** *Agropyron repens* (L.) Beauv.
**Common Name:** triticum
**Other Common Names:** couch grass, dog grass, twitch grass
**Part:** rhizome
**Class:** 1

## *Ephedra distachya* L. Ephedraceae
## *Ephedra equisetina* Bunge.
## *Ephedra gerardiana* Wall. ex Stapf.
## *Ephedra intermedia* Schrenk et C.A. Meyer
## *Ephedra sinica* Stapf.
**Common Name:** ephedra, *ma huang*
**Other Common Names:** Chinese jointfir (*E. sinica*), Indian jointfir (*E. gerardiana*), Chinese ephedra (*E. sinica*)
**Part:** herb
**Class:** 2b [30]; 2c [30]; 2d - Contraindicated in anorexia, bulimia, and glaucoma [4, 21, 22, 24, 28, 29]; thyroid stimulant; not recommended for excessive or long-term use; may potentiate pharmaceutical MAO-inhibitors [30]. Also, see recommended labeling below.

---

Class 3: Herbs for which significant data exist to recommend the following labeling: "To be used only under the supervision of an expert qualified in the appropriate use of this substance." Labeling must include proper use information: dosage, contraindications, potential adverse effects and drug interactions, and any other relevant information related to the safe use of this substance.
Class 4: Herbs for which insufficient data are available for classification.

**Standard Dose:**
Adults: Of the herb: 1.5 to 9.0 grams per day as tea [2, 13, (Yen, 1992)]; of the total alkaloids: 15.0 to 30.0 mg per dose, not to exceed 300 mg per day [4].

Children: Of the total alkaloids: 0.5 mg per kg of body weight, not to exceed 2.0 mg per kilogram body weight per day [4].

**Notice:** MAO Interaction [20, 21] *See page 173.*

Nervous System Stimulant (0.3-2.0% ephedrine alkaloids) [2, 7, 13, 17, 29] *See page 174.*

**Ed. Note:** Excessive use of this herb as a stimulant and for weight loss is widespread. The American Herbal Products Association has adopted and modified recommendations to assure the safe consumption of *Ephedra* since early 1994. The current recommendations are as follows:

(a) to limit daily consumption of total ephedra alkaloids to 120 mg per day in 4 equal doses; and

(b) to use the following cautionary statement on the label of any product containing any species of *Ephedra*, unless the product is documented as free of all ephedra alkaloids, or is in conformity with relevant OTC monographs:

"Seek advice from a health care practitioner prior to use if you are pregnant or nursing, or if you have high blood pressure, heart or thyroid disease, diabetes, difficulty in urination due to prostate enlargement, or if taking a MAO inhibitor or any other prescription drug. Reduce or discontinue use if nervousness, tremor, sleeplessness, loss of appetite, or nausea occur. Not intended for use by persons under 18 years of age. Keep out of the reach of children."

Canadian regulations do not allow ephedra as a non-medicinal ingredient for oral use products (Michols, 1995).

## *Ephedra nevadensis* S. Wats.  Ephedraceae
**Common Name:** Mormon tea
**Other Common Names:** American ephedra, Brigham tea, desert tea
**Part:** herb
**Class:** 1
**Ed. Note:** Because this species contains little or no alkaloids, the recommendation above is not relevant (Osol & Farrar, 1955).

## *Epigaea repens* L.  Ericaceae
**Common Name:** trailing arbutus
**Part:** leaf
**Class:** 1

---

Class 1: Herbs that can be safely consumed when used appropriately.
Class 2: Herbs for which the following use restrictions apply, unless otherwise directed by an expert qualified in the use of the described substance:
    (2a) For external use only;    (2b) Not to be used during pregnancy;
    (2c) Not to be used while nursing;    (2d) Other specific use restrictions as noted.

## *Epilobium angustifolium* L. — Onagraceae
**Common Name:** fireweed
**Other Common Names:** willow herb
**Part:** herb
**Class:** 1


## *Epilobium parviflorum* Schreb. — Onagraceae
**Common Name:** small-flowered willow herb
**Part:** herb
**Class:** 1


## *Epimedium grandiflorum* C. Morr. — Berberidaceae
**Common Name:** epimedium, *yin yang huo*
**Other Common Names:** barrenwort
**Part:** leaf
**Class:** 2d - Not for long-term use [2].
**Ed. Note:** Possible side effects of extended use include dizziness, vomiting, dry mouth, thirst, and nosebleed [2].

In very large doses, Japanese epimedium can cause respiratory arrest and is reported to cause hyperreflexia to the point of mild spasm [2].

*E. brevicornum*, *E. horeanum*, and *E. sagittatum* and as many as 12 other species are interchangeable as *yin yang huo*. Leung & Foster [17] state that *E. grandiflorum* is, in fact, the Japanese species.


## *Equisetum arvense* L. — Equisetaceae
## *Equisetum telmateia* Ehrh.
**Common Name:** horsetail
**Other Common Names:** common horsetail (*E. arvense*), field horsetail (*E. arvense*), shave grass, shavetail grass (*E. arvense*), giant horsetail (*E. telmateia*)
**Part:** herb
**Class:** 2d - *E. arvense* is contraindicated in cardiac or renal dysfunction [4, 5].
**Ed. Note:** Consumption of *Equisetum* is reported to have the potential to lead to thiamine deficiency, and products sold in Canada are required to be certified as free from any thiaminase-like effect [17].

Van Hellemont [24] states that daily use of the powdered extract of the herb should not exceed 2.0 grams and that doses in excess of 5.0 grams a day of the herb powder should be taken during meals.

The herb in powdered form is not recommended for children or for

---

Class 3: Herbs for which significant data exist to recommend the following labeling: "To be used only under the supervision of an expert qualified in the appropriate use of this substance." Labeling must include proper use information: dosage, contraindications, potential adverse effects and drug interactions, and any other relevant information related to the safe use of this substance.
Class 4: Herbs for which insufficient data are available for classification.

prolonged use due to the inorganic silica content [30], though decoctions contain mainly organic silica in colloidal form so are not problematic in this regard [26]. Toxicity is reported to be "...similar to nicotine poisoning...in children who have chewed the stem" [17].

Adulteration of E. arvense with E. palustre, which contains the potentially toxic alkaloid palustrine, is widely reported [5, 18, 27]. Also from Wichtl [27]: "...it has not yet been established whether palustrine-containing drugs are indeed toxic to man. Toxicity data are only available for animals, e.g. cattle, which had eaten large amounts of E. palustre." Other authors report that consumption of large amounts of E. arvense and the related species E. telmateia and E. palustre are poisonous to animals [11, 15, 19, 27].

## Equisetum hyemale L.      Equisetaceae
Common Name: scouring rush, mu zei
Other Common Names: rough horsetail, common scouring rush
Part: herb
Class: 2b [2]

## Eriobotrya japonica (Thunb.) Lindley.      Rosaceae
Common Name: loquat, pi pa ye
Part: leaf
Class: 2d - Not for long-term use; do not exceed recommended dose [30].
Standard Dose: 4.5-12.0 grams of dried leaf or 15.0-30.0 grams of fresh leaf, daily as a tea [2].
Notice: Cyanogenic Glycosides (0.06% amygdalin) [2, 18, 29] *See page 141.*
Ed. Note: The hairs should be removed to avoid irritation of the mucous membranes (Tu, 1988). While material imported from China has, as a rule, been properly processed with regard to this concern, added caution should be exercised when utilizing domestic sources.

## Eriodictyon californicum (Hook. & Arn.) Torr.    Hydrophyllaceae
## Eriodictyon tomentosum Benth.
Common Name: yerba santa (E. californicum)
Other Common Names: woolly yerba santa (E. tomentosum)
Part: herb
Class: 1

---

Class 1: Herbs that can be safely consumed when used appropriately.
Class 2: Herbs for which the following use restrictions apply, unless otherwise directed by an expert qualified in the use of the described substance:
    (2a) For external use only;    (2b) Not to be used during pregnancy;
    (2c) Not to be used while nursing;    (2d) Other specific use restrictions as noted.

*Eryngium maritinum* L.  Apiaceae
*Eryngium planum* L.
*Eryngium yuccifolium* Michx.
Common Name: eryngo, sea holly
Part: root
Class: 1


*Erythrina variegata* L. var. *orientalis* L.  Fabaceae
Synonym: *Erythrina indica* Lam.
Common Name: *hai tong pi*
Part: bark
Class: 1


*Erythroxylum catuaba* A. J. da Silva ex Hamet.  Erythroxylaceae
Common Name: catuaba
Other Common Names: golden trumpet
Part: bark
Class: 4


*Eschscholzia californica* Cham.  Papaveraceae
Common Name: California poppy
Part: whole plant in flower
Class: 2b [4, (Moore, 1993)]; 2d- May potentiate pharmaceutical MAO-inhibitors [30].
Notice: MAO Interaction [17, 28] *See page 173.*
Standard Dose: 2.0-3.5 grams in tea, up to 4 times daily (Moore, 1993).
Ed. Note: The following label has been proposed for products sold in Australia: "Warning: Do not exceed the stated dose" [1].


*Eucalyptus globulus* Labill.  Myrtaceae
Common Name: eucalyptus
Other Common Names: blue gum, Tasmanian blue gum, southern blue gum
Part: leaf
Class: 2d - Contraindicated in inflammatory diseases of the bile ducts and gastrointestinal tract and in severe liver diseases [4]; do not use eucalyptus preparations on areas of the face and especially the nose in infants and young children.
Notice: Tannins (up to 11.0%) [8, 27] *See page 155.*

---

Class 3: Herbs for which significant data exist to recommend the following labeling: "To be used only under the supervision of an expert qualified in the appropriate use of this substance." Labeling must include proper use information: dosage, contraindications, potential adverse effects and drug interactions, and any other relevant information related to the safe use of this substance.
Class 4: Herbs for which insufficient data are available for classification.

## *Eucommia ulmoides* Oliver.            Eucommiaceae
Common Name: eucommia, *du zhong*
Other Common Names: hardy rubber tree
Part: bark
Class: 1


## *Euonymus atropurpureus* Jacq.            Celastraceae
Common Name: wahoo
Other Common Names: eastern burningbush
Part: root bark
Class: 3 [5, 12, 15, 16, 18, 22]


## *Eupatorium perfoliatum* L.            Asteraceae
Common Name: boneset
Other Common Names: thoroughwort
Part: herb
Class: 4
Ed. Note: Although De Smet [9] reports that the alkaloids in this species "...have not been characterized...", Leung & Foster [17] discourage the use of boneset due to the prevalence of toxic pyrrolizidine alkaloids throughout the genus. *E. cannabinum* and *E. purpureum* are known to contain toxic pyrrolizidine alkaloids and should not be substituted for *E. perfoliatum*.
    Large doses are both emetic and cathartic [9, 17, 22].


## *Eupatorium purpureum* L.            Asteraceae
Common Name: Joe Pye
Other Common Names: gravel root, queen-of-the-meadow, Joe-Pye-weed
Part: herb, root, and rhizome
Class: 2a [30]; 2b [8, 9]; 2c [8, 9]; 2d - long-term use is not recommended [30].
Notice: Pyrrolizidine Alkaloids [9] *See page 149.*
Ed. Note: Effective July 1996, the AHPA Board of Trustees recommends that all products with botanical ingredient(s) which contain toxic pyrrolizidine alkaloids, including *Eupatorium purpureum*, display the following cautionary statement on the label:
    "For external use only. Do not apply to broken or abraded skin. Do not use when nursing."

---

Class 1: Herbs that can be safely consumed when used appropriately.
Class 2: Herbs for which the following use restrictions apply, unless otherwise directed by an expert qualified in the use of the described substance:
    (2a) For external use only;     (2b) Not to be used during pregnancy;
    (2c) Not to be used while nursing;     (2d) Other specific use restrictions as noted.

*Euphorbia pilulifera* L.  Euphorbiacea
**Common Name:** euphorbia
**Other Common Names:** pill-bearing spurge
**Part:** herb
**Class:** 2d - May cause nausea and vomiting [10, 18].
**Standard Dose:** As an expectorant: 2.0 grams (Powers et al., 1942; Cook & Martin, 1948); the emetic dose may be similar.
**Notice:** Emetic [10, 18] *See page 167.*
**Ed. Note:** An irritant to the gastrointestinal tract [10].

Canadian regulations do not allow euphorbia in foods [3]. The following label has been recommended in Australia: "Warning: Do not exceed the stated dose" [1].

*Euphrasia officinalis* L.  Scrophulariaceae
**Common Name:** eyebright
**Part:** herb
**Class:** 1

*Euryale ferox* Salib.  Nymphaeaceae
**Common Name:** euryale, *qian shi*
**Other Common Names:** fox nut
**Part:** seed
**Class:** 1

*Evernia furfuracea* (L.) W. Mann.  Usneaceae
*Evernia prunastri* (L.) Achar.
**Common Name:** tree moss (*E. furfuracea*), oak moss (*E. prunastri*)
**Part:** thallus
**Class:** 2d - Not for long-term use; do not exceed recommended dose [30].
**Notice:** Thujone [20, (Furia & Bellanca, 1971)] *See page 158.*
**Ed. Note:** Current U.S. regulations require finished foods which contain this ingredient to be thujone-free [20].

The hot alcohol extract of *Evernia* produces a toxic ethyl ester and is therefore not for internal consumption (Furia & Bellanca, 1971).

---

Class 3: Herbs for which significant data exist to recommend the following labeling: "To be used only under the supervision of an expert qualified in the appropriate use of this substance." Labeling must include proper use information: dosage, contraindications, potential adverse effects and drug interactions, and any other relevant information related to the safe use of this substance.
Class 4: Herbs for which insufficient data are available for classification.

## *Evodia rutaecarpa* J.D. Hook. ex Benth.  Rutaceae
**Common Name:** evodia, *wu zhu yu*
**Part:** fruit
**Class:** 2d - Do not exceed recommended dose [2].
**Standard Dose:** 3.0-9.0 grams daily as tea [2].
**Ed. Note:** Large doses "...have a stimulating effect on the central nervous system and can lead to visual disturbances and hallucinations" [2].

## *Ferula assa-foetida* L.  Apiaceae
## *Ferula foetida* (Bunge) Regel.
## *Ferula rubricaulis* Boiss.
**Common Name:** asafetida
**Other Common Names:** asafoetida, devil's dung, giant fennel
**Part:** oleo gum-resin from rhizomes and roots
**Class:** 2b [10]; 2d - Contraindicated for infant colic [5].
**Notice:** Emmenagogue/Uterine Stimulant [10] *See page 169.*
**Ed. Note:** The classifications and concerns for this herb are based upon therapeutic use and may not be relevant to its consumption as a spice.

Roth [22] reports that 50-100 milligrams of the gum-resin may cause convulsions in persons suffering from nervousness.

Asafetida is commonly used as a food flavoring. Side effects reported for excessive consumption include swollen lips, gastric burning, belching, flatulence and diarrhea, burning urination, headaches, and dizziness [8, 22]. The common source for these concerns is cited as Lewin (Lewin, 1962) who, according to De Smet, "...does not provide an original reference to back up these claims." Nonetheless, De Smet cautions against use of large quantities by persons with peptic ulcers.

Several references cite a single report of methemoglobinemia in a 5-week-old male infant due to glycerited asafetida [5, 21]. While this activity was duplicated *in vitro* with glycerited asafetida, non-glycerited asafetida was negative for the same test and "...had no demonstrable oxidative effect" [8].

## *Filipendula ulmaria* (L.) Maxim.  Rosaceae
**Synonym:** *Spiraea ulmaria* L.
**Common Name:** meadowsweet
**Other Common Names:** queen-of-the-meadow
**Part:** herb
**Class:** 1
**Notice:** Salicylates (Bruneton, 1995) *See page 154.*

---

Class 1: Herbs that can be safely consumed when used appropriately.
Class 2: Herbs for which the following use restrictions apply, unless otherwise directed by an expert qualified in the use of the described substance:
 (2a) For external use only;  (2b) Not to be used during pregnancy;
 (2c) Not to be used while nursing;  (2d) Other specific use restrictions as noted.

*Foeniculum vulgare* Mill.  Apiaceae
**Common Name:** fennel
**Other Common Names:** fennel seed
**Part:** fruit (commonly know as "seed")
**Class:** 1
**Notice:** Estragole (5-10% of essential oil) *See page 143*.
**Ed. Note:** The classifications and concerns for this herb are based upon therapeutic use and may not be relevant to its consumption as a spice.
  Commission E [4] categorizes *F. vulgare* as not for prolonged use without a physician's consultation. Presumably this caution is in reference to therapeutic quantities given as 5.0-7.0 grams of seeds (fruit) daily.

*Forsythia suspensa* (Thunb.) Vahl.  Oleaceae
**Common Name:** forsythia, *lian qiao*
**Other Common Names:** golden bells
**Part:** fruit
**Class:** 2b (18, 22)
**Notice:** Emmenagogue/Uterine Stimulant [18, 22] *See page 169*.

*Fouquieria splendens* Engelm.  Fouquieriaceae
**Common Name:** ocotillo
**Part:** stem
**Class:** 2b (Moore, 1990)

*Fragaria vesca* L.  Rosaceae
*Fragaria virginiana* Duchesne.
**Common Name:** strawberry
**Other Common Names:** alpine strawberry (*F. vesca*), Virginian strawberry (*F. virginiana*)
**Part:** leaf
**Class:** 1

*Fraxinus americana* L.  Oleaceae
*Fraxinus excelsior* L.
**Common Name:** white ash
**Part:** bark
**Class:** 1

---

Class 3: Herbs for which significant data exist to recommend the following labeling: "To be used only under the supervision of an expert qualified in the appropriate use of this substance." Labeling must include proper use information: dosage, contraindications, potential adverse effects and drug interactions, and any other relevant information related to the safe use of this substance.
Class 4: Herbs for which insufficient data are available for classification.

## *Fritillaria cirrhosa* D. Don.                                Liliaceae
## *Fritillaria thunbergii* Miq.
**Common Name:** fritillary, *bei mu*
**Part:** processed bulb
**Class:** 2b [2]
**Standard Dose:** 3.0-15.0 grams daily as tea; 1.0-1.5 grams as powder [2, 13].
**Ed. Note:** Although the unprocessed bulb is toxic [2], commercial sources are generally processed.

The following label has been recommended in Australia for *F. cirrhosa*: "Warning: Do not exceed the stated dose" [1].

Canadian regulations list *F. thunbergii* as an unacceptable non-medicinal ingredient for oral use products (Michols, 1995).

## *Fucus vesiculosus* L.                                                 Fucaceae
**Common Name:** bladderwrack
**Other Common Names:** rockwrack
**Part:** thallus
**Class:** 2b [5]; 2c [5]; 2d - Therapeutic use is not recommended in hyperthyroidism [4, 5, 24, 26]; long-term therapeutic use is not recommended [30].
**Notice:** Iodine (0.02-0.03% dry weight) [27, (Chapman, 1970)] *See page 145.*

## *Galium aparine* L.                                                    Rubiaceae
## *Galium verum* L.
**Common Name:** cleavers
**Other Common Names:** goosegrass (*G. aparine*), lady's bedstraw (*G. verum*), cheese rennet (*G. aparine*, *G. verum*), yellow bedstraw (*G. verum*), our Lady's bedstraw (*G. verum*)
**Part:** herb
**Class:** 1

## *Galium odoratum* (L.) Scop.                                  Rubiaceae
**Synonym:** *Asperula odorata* L.
**Common Name:** sweet woodruff
**Other Common Names:** woodruff
**Part:** herb
**Class:** 1
**Ed. Note:** The use of preparations of sweet woodruff is reported to sometimes result in headaches [22].

Regulated in the U.S. as an allowable flavoring agent in alcoholic beverages only [20].

---

Class 1: Herbs that can be safely consumed when used appropriately.
Class 2: Herbs for which the following use restrictions apply, unless otherwise directed by an expert qualified in the use of the described substance:
    (2a) For external use only;         (2b) Not to be used during pregnancy;
    (2c) Not to be used while nursing;     (2d) Other specific use restrictions as noted.

## *Ganoderma lucidum* (Leyss. ex Fr.) P. Karst.  Ganodermataceae
**Common Name:** reishi, *ling zhi*
**Other Common Names:** *ling chih*, *ling chih* mushroom
**Part:** fruiting body, mycelium
**Class:** 1
**Ed. Note:** Rare side effects (dryness of the mouth, throat, and nasal area; itchiness; stomach upset; nosebleed; bloody stools) have been recorded after 3 to 6 months of continuous use [17]. One instance of skin rash was observed following the consumption of reishi wine [17].

## *Gardenia jasminoides* Ellis.  Rubiaceae
**Common Name:** gardenia fruits, *zhi zi*
**Part:** fruit
**Class:** 1

## *Gastrodia elata* Bl.  Orchidaceae
**Common Name:** gastrodia, *tian ma*
**Part:** rhizome
**Class:** 1

## *Gaultheria procumbens* L.  Ericaceae
**Common Name:** wintergreen
**Other Common Names:** tea berry
**Part:** leaf
**Class:** 1
**Notice:** Salicylates (0.5-1.0%) [28] *See page 154.*
**Ed. Note:** Canadian regulations do not allow wintergreen as a non-medicinal ingredient for oral use products, except as a flavor (Michols, 1995).

## *Gelidiella acerosa* (Forssk.) Feldm.  Gelidiaceae
**Common Name:** agar
**Other Common Names:** agarweed
**Part:** thallus
**Class:** 2d - Take with at least 250 ml (8 oz) liquid [20]; contraindicated in bowel obstruction [30]
**Standard Dose:** 4.0-16.0 grams, one to two times daily [21].
**Notice:** Bulk-forming laxative [30] *See page 165.*
**Ed. Note:** Specific labeling is required in the U.S. for all OTC drug products containing agar [20] *See Appendix 2, page 165.*

---

Class 3: Herbs for which significant data exist to recommend the following labeling: "To be used only under the supervision of an expert qualified in the appropriate use of this substance." Labeling must include proper use information: dosage, contraindications, potential adverse effects and drug interactions, and any other relevant information related to the safe use of this substance.

Class 4: Herbs for which insufficient data are available for classification.

*Gelidium amansii* J. V. Lamour.  Gelidiaceae
*Gelidium cartilagineum* (L.) Gaill.
*Gelidium crinale* (Turn.) J. V. Lamour.
*Gelidium divaricatum* G. Martens.
*Gelidium pacificum* Okam.
*Gelidium vagum* Okam.
**Common Name:** agar
**Other Common Names:** agarweed
**Part:** thallus
**Class:** 2d - Take with at least 250 ml (8 oz) liquid [20]; contraindicated in bowel obstruction [30].
**Standard Dose:** 4.0-16.0 grams, one to two times daily [21].
**Notice:** Bulk-forming laxative [17, 21, 28] *See page 165.*
**Ed. Note:** Specific labeling is required in the U.S. for all OTC drug products containing agar [20] *See Appendix 2, page 165.*

*Genista tinctoria* L.  Fabaceae
**Common Name:** dyer's broom
**Other Common Names:** dyer's greenwood, broom flower
**Part:** herb and flower
**Class:** 2b [30]; 2d - May cause nausea and vomiting [28].
**Notice:** Emetic [28] *See page 167.*
**Ed. Note:** Although Roth [22] lists dyer's broom as "toxic", this is not substantiated by references 18, 27, or 28 or by Madaus (1976).

Constituents found in *G. tinctoria* are similar to the alkaloids in *Cytisus scoparius* which are classified as uterine stimulants [30].

*Gentiana lutea* L.  Gentianaceae
**Common Name:** gentian
**Other Common Names:** yellow gentian, wild gentian
**Part:** root
**Class:** 2d - Contraindicated in gastric and duodenal ulcers [4, 5] and when gastric irritation and inflammation are present [10, 24].
**Ed. Note:** Irritating qualities are maximized in tincture form and minimized as a tea [30].

The German Standard License label states headaches may occasionally occur in persons sensitive to bitter substances [4].

Leung & Foster [17] report that gentian may not be well tolerated by pregnant women or persons with high blood pressure, both conditions for which their reference (Tyler, 1993) advises caution "about using any medication, herbal or otherwise."

---

Class 1: Herbs that can be safely consumed when used appropriately.
Class 2: Herbs for which the following use restrictions apply, unless otherwise directed by an expert qualified in the use of the described substance:
   (2a) For external use only;   (2b) Not to be used during pregnancy;
   (2c) Not to be used while nursing;   (2d) Other specific use restrictions as noted.

## *Gentiana macrophylla* Bge.    Gentianaceae
**Common Name:** large-leaf gentian, *qin jiao*
**Part:** root
**Class:** 1
**Ed. Note:** High doses may cause nausea and vomiting [2].
The concerns stated for G. *lutea*, though not specified in available references, may also be relevant to this species [30].

## *Gentiana scabra* Bge.    Gentianaceae
**Common Name:** scabrous gentian, *long dan*
**Part:** root
**Class:** 1
**Ed. Note:** When administered after meals or in excessive dosage, G. *scabra* may cause impairment of the digestive function and occasionally, headache, flushing of the face, and vertigo [7].

The concerns stated for G. *lutea*, though not specified in available references, may also be relevant to this species [30].

## *Geranium maculatum* L.    Geraniaceae
**Common Name:** cranesbill
**Other Common Names:** wild geranium, alumroot
**Part:** root
**Class:** 1

## *Ginkgo biloba* L.    Ginkgoaceae
**Common Name:** ginkgo
**Other Common Names:** maidenhair tree
**Part:** leaf
**Class:** 2d- May potentiate pharmaceutical MAO-inhibitors [30].
**Notice:** MAO Interaction [17, 28] *See page 173*.
**Ed. Note:** Ginkgo leaf preparations are generally sold either as a tincture, for which no side effects have been recorded, or as a highly concentrated extract. Ginkgo leaf extracts (concentrated to 24.0% flavone glycosides and 6.0% terpenes) are among the most actively studied of the modern phytopharmaceutical substances. With over 5 million prescriptions per year recorded in Germany alone, only occasional side effects, such as headaches and gastrointestinal upset, have been reported (Hobbs, 1991b). A review of all spontaneous reports from 1982 to 1988 of adverse events associated with the most established concentrated extract concluded that such side effects are rare, and that "...tolerance was generally excellent" (DeFeudis, 1991).

---

Class 3: Herbs for which significant data exist to recommend the following labeling: "To be used only under the supervision of an expert qualified in the appropriate use of this substance." Labeling must include proper use information: dosage, contraindications, potential adverse effects and drug interactions, and any other relevant information related to the safe use of this substance.

Class 4: Herbs for which insufficient data are available for classification.

**Part:** seed
**Class:** 2d - Do not exceed recommended dose; not for long-term use [2, 7, 17].
**Standard Dose:** Prepared: 4.5-15.0 grams daily [2, 13]; raw: "dosage is reduced", quantity not given [2].
**Ed. Note:** Various references are inconsistent in addressing the toxicity of ginkgo seeds. Comments range from: "Because of its higher toxicity, when the raw herb [seed] is used the dosage is reduced" [2] to: "Fresh seeds are toxic and have been reported to cause death in children" [17]. Safe consumption of the cooked seeds is well documented [6, 17], but Leung & Foster [17] advise a limit of no more than 10 seeds a day for the boiled or roasted seeds.

Canadian regulations do not allow ginkgo seeds in foods (Welsh, 1995).

The pulpy fruit present in the fresh state may cause contact dermatitis, GI upset, and other unpleasant effects [7, 17].

### *Glehnia littoralis* Fr. Schmidt ex Miq.   Apiaceae
**Common Name:** glehnia, beach silvertop, *bei sha shen*
**Part:** root
**Class:** 1

### *Glycyrrhiza glabra* L.   Fabaceae
### *Glycyrrhiza echinata* L.
### *Glycyrrhiza uralensis* Fisch ex DC.
**Common Name:** licorice, *gan cao*
**Other Common Names:** liquorice
**Part:** root
**Class:** 2b [4, 5, 6, 24, 25]; 2c [24]; 2d - Not for prolonged use or in high doses except under supervision of a qualified health practitioner [2, 4, 5, 6, 14, 21, 24]; contraindicated for diabetics [24] and in hypertension, liver disorders, severe kidney insufficiency, and hypokalemia [4, 5, 9]; may potentiate potassium depletion of thiazide diuretics and stimulant laxatives [4], as well as the action of cardiac glycosides [4, 24] and cortisol [4, 9, 14].
**Standard Dose:** 1.0-5.0 grams, three times daily for up to 6 weeks, though French regulation limits daily consumption to 5.0 grams for direct consumption or 8.0 grams as a tea; reduction in sodium and increase in potassium intakes are recommended [4, 9, 24, (Mitchell et al., 1983)].
**Ed. Note:** May cause reversible potassium depletion and sodium retention, resulting in such symptoms as hypertension, edema, headache, and vertigo when consumed in therapeutic dosages over a prolonged period [2, 4, 5, 14, 21, 24].

Deglycyrrhizinised licorice is usually free of adverse effects [21].

---

Class 1: Herbs that can be safely consumed when used appropriately.
Class 2: Herbs for which the following use restrictions apply, unless otherwise directed by an expert qualified in the use of the described substance:
   (2a) For external use only;   (2b) Not to be used during pregnancy;
   (2c) Not to be used while nursing;   (2d) Other specific use restrictions as noted.

*Gossypium herbaceum* L.  Malvaceae
*Gossypium hirsutum* L.
**Common Name:** cotton
**Other Common Names:** cotton root bark (G. *hirsutum*), levant cotton (G. *herbaceum*)
**Part:** root bark
**Class:** 2b [1, 3, 6, 18]; 2d - Contraindicated in urogenital irritation or tendency to inflammation [10]; chronic use may cause sterility in men [9, 28].
**Notice:** Abortifacient [6, 9, 10] *See page 163.*
Emmenagogue/Uterine Stimulant [6, 10, 18, 28] *See page 169.*
**Ed. Note:** Current Canadian regulations do not allow an excess of 450 ppm of free gossypol in foods [3]. This is especially relevant to such food products as cotton seed meal and oil.

*Grifola frondosa* (Fr.) S. F. Gray  Polyporaceae
**Common Name:** maitake
**Other Common Names:** hen of the woods
**Part:** fruiting body, mycelium
**Class:** 1

*Grifola umbellata* (Pers. ex Fr.)  Polyporaceae
**Synonym:** Polyporus umbellatus (Pers.) Fr.
**Common Name:** *zhu ling*
**Other Common Names:** grifola, polyporus
**Part:** fruiting body, mycelium
**Class:** 1

*Grindelia robusta* Nutt.  Asteraceae
*Grindelia squarrosa* (Pursh) Dunal.
**Common Name:** grindelia
**Other Common Names:** gumweed
**Part:** herb
**Class:** 1
**Ed. Note:** High doses can produce kidneys and stomach irritation [4, 24].

*Hamamelis virginiana* L.  Hamamelidaceae
**Common Name:** witch-hazel
**Part:** bark and leaf
**Class:** 1

---

Class 3: Herbs for which significant data exist to recommend the following labeling: "To be used only under the supervision of an expert qualified in the appropriate use of this substance." Labeling must include proper use information: dosage, contraindications, potential adverse effects and drug interactions, and any other relevant information related to the safe use of this substance.
Class 4: Herbs for which insufficient data are available for classification.

**Notice:** Tannins (10.0% bark, 3-10.0% leaf) [17, 27] *See page 155.*
**Ed. Note:** "In susceptible patients, irritation of the stomach may occur occasionally. In rare cases, witch-hazel tannins may cause liver damage" [27]. It should be noted that witch-hazel water is a steam distillate and does not contain tannins [17].

### *Harpagophytum procumbens* DC. ex Meisn. — Pedaliaceae
**Common Name:** devil's claw
**Part:** secondary tuber
**Class:** 2d - Contraindicated in gastric and duodenal ulcers [4, 5, 24].

### *Hedeoma pulegioides* (L.) Pers. — Lamiaceae
**Common Name:** American pennyroyal
**Other Common Names:** pennyroyal
**Part:** herb
**Class:** 2b [10]
**Ed. Note:** *H. pulegoides* and *Mentha pulegium* are historically interchangeable as the source for pennyroyal oil [8]. Although the American species is reported to contain less of the toxic ketone, pulegone, than the European plant (Furia & Belanca, 1971), the cautions presented in the listing for *Mentha pulegium* are relevant to American pennyroyal.

Canadian regulations do not allow American pennyroyal in foods except in alcoholic beverages if pulegone-free [3].

### *Helianthus annuus* L. — Asteraceae
**Common Name:** sunflower
**Part:** seed
**Class:** 1

### *Hemidesmus indicus* (L.) Schult. — Asclepiadaceae
**Common Name:** hemidesmus
**Other Common Names:** Indian sarsaparilla, East Indian sarsaparilla
**Part:** root
**Class:** 4

---

Class 1: Herbs that can be safely consumed when used appropriately.
Class 2: Herbs for which the following use restrictions apply, unless otherwise directed by an expert qualified in the use of the described substance:
    (2a) For external use only;    (2b) Not to be used during pregnancy;
    (2c) Not to be used while nursing;    (2d) Other specific use restrictions as noted.

## *Hepatica nobilis* Garsault var. *acuta* (Pursh) Steyermark.
## *Hepatica nobilis* Garsault var. *obtusa* (Pursh) Steyermark.

Ranunculaceae

**Common Name:** American liverleaf
**Other Common Names:** liverwort herb
**Part:** herb
**Class:** 2b [4]
**Ed. Note:** High doses may irritate the kidneys and urinary tract [4].

*Hepatica*, as with several species of the Ranunculaceae, contains protoanemonin, a strongly irritating vesicant oil; however, solubility in water is only 1% (Windholz, 1983). The herb can cause subepidermal blistering of the skin in its fresh state, though the compound and the related concern are destroyed upon drying [4, 18, (Mitchell & Rook, 1979)].

## *Heuchera micrantha* Douglas ex Lindl.

Saxifragaceae

**Common Name:** alumroot
**Part:** root
**Class:** 1
**Notice:** Tannins (dried root: 9.3-19.6%) [Osol & Farrar, 1955] *See page 155.*

## *Hibiscus sabdariffa* L.

Malvaceae

**Common Name:** hibiscus
**Other Common Names:** roselle, red sorrel
**Part:** flowers
**Class:** 1
**Ed. Note:** Although regulated in the U.S. as an allowable flavoring agent in alcoholic beverages only [20], the extensive history of use attributed to hibiscus brings this regulatory limitation into question.

## *Hordeum vulgare* L.

Poaceae

**Common Name:** barley, *mai ya* (germinated barley)
**Part:** sprouted seed
**Class:** 2b [7, 13]

## *Humulus lupulus* L.

Cannabinaceae

**Common Name:** hops
**Part:** strobiles
**Class:** 2d - Some writers advise against use in depression [5, (Mitchell et al., 1983)].

---

Class 3: Herbs for which significant data exist to recommend the following labeling: "To be used only under the supervision of an expert qualified in the appropriate use of this substance." Labeling must include proper use information: dosage, contraindications, potential adverse effects and drug interactions, and any other relevant information related to the safe use of this substance.
Class 4: Herbs for which insufficient data are available for classification.

### *Hydrangea arborescens* L. — Hydrangeaceae
**Common Name:** hydrangea
**Other Common Names:** wild hydrangea
**Part:** root
**Class:** 2d - Not for long-term use; do not exceed recommended dose [30].
**Standard Dose:** 2.0 grams (Powers et al., 1942).
**Notice:** Cyanogenic Glycosides (hydrangin, about 1.0%) [12, (Cook and Martin, 1948)] *See page 141.*
**Ed. Note:** Flowers and leaves are reported to have caused toxic symptoms in humans. This is assumed to be due to the cyanogenic glycoside hydrangin, which is also reported to be found in the root [16, 17, 18].

### *Hydrastis canadensis* L. — Ranunculaceae
**Common Name:** goldenseal
**Other Common Names:** yellow puccoon, orangeroot
**Part:** rhizome and root
**Class:** 2b [1, 3, 5, 18, (Welsh, 1995)]
**Standard Dose:** Unless otherwise prescribed, three times daily: 500-1000 mg dried rhizome and root; 2.0-4.0 ml tincture (1:10, 60% ethanol); 0.3-1.0 ml liquid extract (1:1, 60% ethanol) [5, 28].
**Notice:** Berberine (0.5-6.0%) [5, 8, 10, 21, 28] *See page 136.*
Emmenagogue/Uterine Stimulant [18] *See page 169.*
**Ed. Note:** Fresh plant may cause irritation to the mucosa [19].

Canadian regulations do not allow goldenseal as a non-medicinal ingredient for oral use products (Michols, 1995).

### *Hypericum perforatum* L. — Hypericaceae
**Common Name:** St. John's wort
**Part:** herb, flowering tops
**Class:** 2d - May potentiate pharmaceutical MAO-inhibitors [30].
**Notice:** MAO Interaction [17, 28] *See page 173.*
Photosensitizing [4, 17, 19, 21, 24, 26, 27] *See page 176.*
Tannins (up to 10.0%) [17, 27] *See page 155.*
**Ed. Note:** Although phototoxicity in human beings [17, 27] is rare, fair-skinned individuals should avoid excessive exposure to sunlight when using *H. perforatum* [4, 17, 19, 21, 24, 26, 27]. A standardized *Hypericum* extract (600 mg 3x daily yielding 0.24-0.32% total hypericin) produced a measurable increase in erythema in light-sensitive volunteers when exposed to uv light (Cott, 1996).

*H. perforatum* is regulated in the U.S. as an allowable flavoring agent in

---

Class 1: Herbs that can be safely consumed when used appropriately.
Class 2: Herbs for which the following use restrictions apply, unless otherwise directed by an expert qualified in the use of the described substance:
 (2a) For external use only; (2b) Not to be used during pregnancy;
 (2c) Not to be used while nursing; (2d) Other specific use restrictions as noted.

alcoholic beverages only, with a further limitation that only the hypericin-free alcohol distillate form is allowed [20].

### *Hyssopus officinalis* L. — Lamiaceae
**Common Name:** hyssop
**Part:** herb
**Class:** 2b (Madaus, 1976)
**Notice:** Emmenagogue/Uterine Stimulant [30] *See page 169*.

### *Ilex paraguayensis* St. Hil. — Aquifoliaceae
**Common Name:** maté
**Other Common Names:** Paraguay tea, yerba maté
**Part:** leaf
**Class:** 2d - Not recommended for excessive or long-term use [30].
**Notice:** Nervous System Stimulant (0.3-1.7% caffeine) [18, 27] *See page 174*. Tannins (4.0-16.0%) [18, 27] *See page 155*.

### *Illicium verum* J. D. Hook. — Illiciaceae
**Common Name:** star anise
**Part:** fruit
**Class:** 1
**Ed. Note:** *I. verum* should not be confused with the toxic Japanese, or bastard star anise, *I. lanceolatum*, or *I. anisatum* (syn. *I. religiosum*) [14, 27, 28], though such adulteration is rare [28]. The fruit of the Japanese variety is generally smaller than that of *I. verum* and consists of 6-8 follicles arranged in a star pattern and terminating in an upward curing tip. This curve is in contrast to the nearly straight beak of the standard species. Japanese star anise is reported to have a bitter flavor not found in the star anise of commerce [28, (Youngken, 1921)].

### *Imperata cylindrica* L. — Poaceae
**Common Name:** woolly grass, *mao gen*
**Part:** rhizome
**Class:** 1

### *Inula britannica* L. — Asteraceae
**Common Name:** British elecampane, *xuan fu hua*
**Part:** flower
**Class:** 1

---

Class 3: Herbs for which significant data exist to recommend the following labeling: "To be used only under the supervision of an expert qualified in the appropriate use of this substance." Labeling must include proper use information: dosage, contraindications, potential adverse effects and drug interactions, and any other relevant information related to the safe use of this substance.
Class 4: Herbs for which insufficient data are available for classification.

## *Inula helenium* L. — Asteraceae
Common Name: elecampane
Other Common Names: scabwort, alant, horseheal, yellow starwort
Part: rhizome and root
Class: 2b [5]; 2c [5]
Ed. Note: Large dosage causes vomiting, diarrhea, spasms, and symptoms of paralysis [4, 22, 27].

Regulated in the U.S. as an allowable flavoring agent in alcoholic beverages only [20].

## *Ipomoea purga* Hayne. — Convolvulaceae
Common Name: jalap
Part: root
Class: 3 [8, 10, 17, 21, 24]

## *Iris versicolor* L. — Iridaceae
## *Iris virginica* L.
Common Name: blue flag
Other Common Names: sweet flag
Part: rhizome and root
Class: 2b [30]; 2d - May cause nausea and vomiting [18].
Notice: Emetic [18] *See page 167.*
Ed. Note: Fresh root may cause irritation to the mucosa [10].

## *Iris germanica* L. — Iridaceae
## *Iris germanica* L. var. *florentina* Dykes.
## *Iris pallida* Lam.
Common Name: orris
Part: root
Class: 1
Ed. Note: Fresh root may cause irritation to the mucosa [6, 10, 24].

## *Isatis tinctoria* L. — Brassicaceae
Common Name: Dyer's-woad (root: *ban lan gen*; leaf: *da qing ye*)
Other Common Names: woad
Part: root
Class: 1

---

Class 1: Herbs that can be safely consumed when used appropriately.
Class 2: Herbs for which the following use restrictions apply, unless otherwise directed by an expert qualified in the use of the described substance:
    (2a) For external use only;    (2b) Not to be used during pregnancy;
    (2c) Not to be used while nursing;    (2d) Other specific use restrictions as noted.

## *Jasminum grandiflorum* L.             Oleaceae
**Common Name:** jasmine
**Other Common Names:** Catalonian jasmine, royal jasmine, Spanish jasmine
**Part:** flower
**Class:** 1

## *Juglans cinerea* L.             Juglandaceae
**Common Name:** butternut
**Part:** bark
**Class:** 1
**Ed. Note:** Large doses can be mildly cathartic and so may be contraindicated in pregnancy [10, 30].

## *Juglans nigra* L.             Juglandaceae
**Common Name:** black walnut
**Part:** hull, leaf
**Class:** 2d - Prolonged use is not advised due to the presence of significant quantities of juglone [18], a known mutagen in animals [4]. Carcinogenic effects associated with the chronic external use of *J. regia* in humans have been observed [4], which may or may not apply to *J. nigra*.

## *Juniperus communis* L.             Cupressaceae
## *Juniperus oxycedrus* L.
**Common Name:** juniper
**Other Common Names:** common juniper
**Part:** fruit (berry)
**Class:** 2b [4, 24, 26, 27, 28]; 2d - Not for use exceeding four to six weeks in succession [9, 24, 26]; contraindicated in inflammatory kidney disease [4, 9, 24, 26, 27].
**Ed. Note:** It is commonly reported that juniper berries should not be used long-term. However, Hiel and Schilcher found that high doses of ripe juniper berries and the essential oil distilled only from the ripe berries can be used safely. The researchers nonetheless recommend that "...care should still be exercised in cases of acute renal inflammation" (Bone, 1995).

When purchasing juniper berries in trade, adulteration with other species is possible. *J. communis* is the preferred species. Information on other species is lacking in our primary references.

Canadian regulations do not allow juniper as a non-medicinal ingredient for oral use products (Michols, 1995).

---

Class 3: Herbs for which significant data exist to recommend the following labeling: "To be used only under the supervision of an expert qualified in the appropriate use of this substance." Labeling must include proper use information: dosage, contraindications, potential adverse effects and drug interactions, and any other relevant information related to the safe use of this substance.
Class 4: Herbs for which insufficient data are available for classification.

*Juniperus monosperma* (Engelm.) Sarg.  Cupressaceae
*Juniperus osteosperma* (Torr.) Little.
**Common Name:** oneseed juniper
**Other Common Names:** juniper
**Part:** fruit
**Class:** 2b [30]
**Ed. Note:** None of our references mention these North American species. We assume the data and classifications which apply to the fruit of *J. communis* and *J. oxycedrus* are applicable.

*Juniperus virginiana* L.  Cupressaceae
**Common Name:** eastern red cedar
**Other Common Names:** cedarwood
**Part:** leaf, berry
**Class:** 2b [30]; 3 [15, 22]

*Kaempferia galanga* L.  Zingiberaceae
**Common Name:** greater galangal
**Part:** root
**Class:** 1

*Krameria argentea* Mart. ex Spreng.  Krameriaceae
*Krameria triandra* Ruiz. & Pav
**Common Name:** rhatany
**Other Common Names:** Brazilian rhatany, Peruvian rhatany, Krameria
**Part:** root
**Class:** 1
**Notice:** Tannins (10.0-15.0% in *Krameria triandra*) [4, 27] *See page 155.*
**Ed. Note:** Undiluted tincture may cause local irritation [9, 27].
    Commission E [4] requires labeling which limits use to 2 weeks without a physician consultation.

*Lactuca quercina* L.  Asteraceae
*Lactuca serriola* L.
*Lactuca virosa* L.
**Common Name:** wild lettuce
**Part:** herb
**Class:** 1

---

Class 1: Herbs that can be safely consumed when used appropriately.
Class 2: Herbs for which the following use restrictions apply, unless otherwise directed by an expert qualified in the use of the described substance:
    (2a) For external use only;    (2b) Not to be used during pregnancy;
    (2c) Not to be used while nursing;   (2d) Other specific use restrictions as noted.

**Standard Dose:** 0.5-4.0 grams, eaten or infused [5].
**Ed. Note:** According to Van Hellemont [24], *Lactuca* is contraindicated in glaucoma and prostate enlargement.

The following label has been recommended in Australia for *Lactuca virosa*: "Warning: Do not exceed the stated dose" [1].

## *Lamium album* L. — Lamiaceae
**Common Name:** white nettle
**Other Common Names:** dead nettle
**Part:** herb
**Class:** 1

## *Larrea tridentata* (Sesse. & Moc. ex DC.) Coville. — Zygophyllaceae
**Common Name:** chaparral
**Other Common Names:** creosote bush
**Part:** leaf
**Class:** 2d - Not for use in large amounts by persons with pre-existing kidney disease [9] and liver conditions, such as hepatitis and cirrhosis [30].
**Ed. Note:** De Smet [9] states that no animal evidence for hepatotoxicity exists in the literature. However, reports of acute liver toxicity associated with consumption of *L. tridentata* surfaced from 1990 through late 1992, leading to the issuance of a warning by the FDA to cease consumption of chaparral [17]. The American Herbal Products Association (AHPA) initiated a review of four cases (Watt et al., 1994) and found the reported toxicity to be due to idiosyncratic reactions in persons with pre-existing liver conditions. The authors concluded, and AHPA recommended in February, 1995, that products containing *L. tridentata* should be labeled with the following cautionary statement:

"Seek advice from a health care practitioner before use if you have any history of liver disease. Discontinue use if nausea, fever, fatigue, or jaundice occur (e.g., dark urine or yellow discoloration of the eyes)."

In addition, the AHPA established a reporting mechanism to track any future concerns. To report unusual conditions, call (301) 951-3204.

Hematological concerns have been raised based on isolated incidents of consumption of both chaparral and its phenolic component, nordihydroguaiaretic acid [9].

Canadian regulations do not allow chaparral as a non-medicinal ingredient for oral use products (Michols, 1995).

---

Class 3: Herbs for which significant data exist to recommend the following labeling: "To be used only under the supervision of an expert qualified in the appropriate use of this substance." Labeling must include proper use information: dosage, contraindications, potential adverse effects and drug interactions, and any other relevant information related to the safe use of this substance.
Class 4: Herbs for which insufficient data are available for classification.

***Laurus nobilis* L.**                                Lauraceae
Common Name: bay leaf
Other Common Names: Grecian laurel, sweet bay
Part: leaf
Class: 1

***Lavandula angustifolia* Mill.**                   Lamiaceae
***Lavandula latifolia* Medik.**
***Lavandula stoechas* L.**
***Lavandula* x *intermedia* Emeric ex Loisel.**
Common Name: lavender
Other Common Names: common lavender, French lavender (*L. angustifolia*), English lavender (*L. angustifolia*), Spanish lavender (*L. stoechas*), spike lavender (*L. latifolia*)
Part: flower
Class: 1

***Lawsonia inermis* L.**                              Lythraceae
Common Name: henna
Part: leaf
Class: 2a [17, 18]
Ed. Note: History of internal use as an abortifacient is recorded in Africa [18].

***Ledebouriella seseloides* (Hoffm.) Wolff.**       Apiaceae
Common Name: siler, *fang feng*
Part: root
Class: 1

***Lentinus edodes* (Berk.) Singer.**             Polyporaceae
Common Name: shiitake
Other Common Names: shiitake mushroom
Part: fruiting body and mycelium
Class: 1

***Leonurus cardiaca* L.**                            Lamiaceae
Common Name: motherwort
Other Common Names: common motherwort
Part: herb

---

Class 1: Herbs that can be safely consumed when used appropriately.
Class 2: Herbs for which the following use restrictions apply, unless otherwise directed by an expert qualified in the use of the described substance:
    (2a) For external use only;     (2b) Not to be used during pregnancy;
    (2c) Not to be used while nursing;  (2d) Other specific use restrictions as noted.

Class: 2b [5, 6, 18, 30]
Notice: Emmenagogue/Uterine Stimulant [6, 18] *See page 169.*
Ed. Note: Van Hellemont [24] states that a dose in excess of 3.0 grams of a powdered extract may cause diarrhea, uterine bleeding, and stomach irritation. No indication of the extract's concentration is recorded. No other reference duplicates these concerns.

## *Leonurus heterophyllus* Sweet   Lamiaceae
## *Leonurus sibiricus* L.
Synonym: *L. artemisia* auct. non Lour.
Common Name: Chinese motherwort, *yi mu cao* (herb), *chong wei zi* (fruit)
Part: fruit
Class: 2b [30]

Part: herb
Class: 2b [2, 7, 23, 30]
Notice: Emmenagogue/Uterine Stimulant [2, 7, 14, 23] *See page 169.*

## *Leptandra virginica* (L.) Nutt.   Scrophulariaceae
Common Name: culver's-root
Other Common Names: black root
Part: root
Class: 1

Part: fresh root
Class: 2b [10]; 2d - The fresh root is violently cathartic [10, 19].

## *Levisticum officinale* W. Koch.   Apiaceae
Common Name: lovage
Part: root
Class: 2b [17, 18]; 2d - Contraindicated in impaired kidney function or inflammation of the kidneys [4, 27].
Notice: Emmenagogue/Uterine Stimulant [17, 18] *See page 169.*
Ed. Note: The classifications and concerns for this herb are based upon therapeutic use and may not be relevant to its consumption as a spice.

Due to photosensitizing potential, caution should be observed in exposure to excessive sunlight during long-term use [4, 17]. However, Wichtl [27] states that "...there is no fear of phototoxic...effects on therapeutic use."

---

Class 3: Herbs for which significant data exist to recommend the following labeling: "To be used only under the supervision of an expert qualified in the appropriate use of this substance." Labeling must include proper use information: dosage, contraindications, potential adverse effects and drug interactions, and any other relevant information related to the safe use of this substance.
Class 4: Herbs for which insufficient data are available for classification.

### *Ligusticum chuanxiong* Hort. — Apiaceae
Synonym: *Ligusticum wallichii* auct. sin. Non. Franch.
Common Name: Sichuan lovage, *chuan xiong*
Other Common Names: Szechuan lovage
Part: rhizome
Class: 2b [2, 7, 14, 29]

### *Ligusticum porteri* J. M. Coulter & J. N. Rose — Apiaceae
Common Name: osha
Part: root
Class: 2b [30]
Ed. Note: Pacific coast species *L. apiifolium* and *L. californicum* are commonly substituted for *L. porteri*.

### *Ligusticum sinense* Lour. — Apiaceae
Common Name: ligusticum, *gao ben*
Part: root and rhizome
Class: 1

### *Ligustrum lucidum* W.T. Aiton. — Oleaceae
Common Name: ligustrum fruits
Other Common Names: glossy privet, *nu zhen zi*
Part: fruit
Class: 1

### *Lilium brownii* F.E. Br. — Liliaceae
Common Name: Brown's lily, *bai he*
Part: bulb
Class: 1

### *Linum usitatissimum* L. — Linaceae
Common Name: flax
Other Common Names: linseed
Part: seed
Class: 2d - Take with at least 150 ml (6 oz) liquid [4, 27]; contraindicated in bowel obstruction [4, 27].
Standard Dose: 10.0 grams (ca. 2½ tsp) whole or cracked seeds with meals, 2-3 times daily [4, 27].
Notice: Bulk-forming Laxative [4, 27] *See page 165*.

---

Class 1: Herbs that can be safely consumed when used appropriately.
Class 2: Herbs for which the following use restrictions apply, unless otherwise directed by an expert qualified in the use of the described substance:
    (2a) For external use only;    (2b) Not to be used during pregnancy;
    (2c) Not to be used while nursing;    (2d) Other specific use restrictions as noted.

**Ed. Note:** The seeds should be preswollen with fluids prior to consumption in cases of bowel inflammation [27].

The risk of reduction in the absorption of other drugs... "as with any other mucilage...", is reported [4].

Only the whole seed should be used by overweight people to avoid caloric absorption [27].

Van Hellemont [24], without reference, warns against use in thyroid insufficiency.

## *Lobelia inflata* L.  Campanulaceae
**Common Name:** lobelia
**Other Common Names:** Indian tobacco, puke weed
**Part:** herb
**Class:** 2b [5]; 2d - May cause nausea and vomiting [5, 10, 12, 17, 18, 28]; not to be taken in large doses [17, 21]; dose-dependent cardioactivity has been observed [21].
**Standard Dose:** As an expectorant: 100 mg of the leaf; 0.6-2.0 ml of the tincture [1, 12, 17, 19, 21, (Osol & Farrar, 1955)].
**Notice:** Emetic [5, 10, 12, 17, 18, 28] *See page 167.*
**Ed. Note:** Controvery over the safety of lobelia dates to the early 19th century. Felter & Lloyd [10] state that the "...emetic action is so prompt and decided, that the contained alkaloid could not, under ordinary circumstances, produce fatal results." We find no substantiated evidence of severe symptoms or death produced by *L. inflata* to contradict this statement.

Canadian regulations do not allow lobelia in foods [3].

## *Lobelia siphilitica* L.  Campanulaceae
**Common Name:** blue lobelia
**Other Common Names:** great blue lobelia
**Part:** herb
**Class:** 2b [5]; 2d - Not to be taken in large doses [17, 19, 21].
**Ed. Note:** Contains similar alkaloids to *L. inflata*, though is lacking in lobeline See Editor's Note for *L. inflata*.

## *Lomatium dissectum* (Nutt.) Math. & Const.  Apiaceae
**Common Name:** lomatium
**Part:** root
**Class:** 2b [30]; 2d - When used internally, *Lomatium* can cause skin rashes [30, (Moore, 1993)].

---

Class 3: Herbs for which significant data exist to recommend the following labeling: "To be used only under the supervision of an expert qualified in the appropriate use of this substance." Labeling must include proper use information: dosage, contraindications, potential adverse effects and drug interactions, and any other relevant information related to the safe use of this substance.
Class 4: Herbs for which insufficient data are available for classification.

### *Lonicera japonica* Thunb.      Caprifoliaceae
**Common Name:** Japanese honeysuckle, *jin yin hua* (flowers), *ren dong teng* (stem)
**Part:** flower, stem
**Class:** 1

### *Lycium barbarum* L.      Solanaceae
### *Lycium chinense* Mill.
**Common Name:** lycium, *gou qi zi* (berry), *di gu pi* (root bark)
**Other Common Names:** matrimony vine, wolfberry
**Part:** berry and root bark
**Class:** 2b [14, 23]

### *Lycopus americanus* Muhl. ex W. Barton.      Lamiaceae
### *Lycopus europaeus* L.
### *Lycopus virginicus* L.
**Common Name:** bugleweed (*L. virginicus*)
**Other Common Names:** water horehound (*L. europaeus, L. americanus*)
**Part:** herb
**Class:** 2b [9]; 2c [9]; 2d - Contraindicated in thyroid enlargement or hypothyroid [4, 6] and in simultaneous administration of other thyroid treatments [4].

### *Magnolia liliflora* Desr.      Magnoliaceae
**Common Name:** red magnolia, *xin yi*
**Part:** flower
**Class:** 1

### *Magnolia officinalis* Rehd. et Wils.      Magnoliaceae
**Common Name:** magnolia, *hou po* (bark), *hou po hua* (flower)
**Part:** bark
**Class:** 2b [2, 29]

**Part:** flower
**Class:** 1
**Ed. Note:** Although Leung and Foster [17] report uterine stimulation in animals given extract or decoction of *Magnolia* flowers, no reference contraindicates use in pregnancy.

---

Class 1: Herbs that can be safely consumed when used appropriately.
Class 2: Herbs for which the following use restrictions apply, unless otherwise directed by an expert qualified in the use of the described substance:
    (2a) For external use only;      (2b) Not to be used during pregnancy;
    (2c) Not to be used while nursing;      (2d) Other specific use restrictions as noted.

Canadian regulations do not allow magnolia as a non-medicinal ingredient for oral use products (Michols, 1995).

### *Magnolia virginiana* L. — Magnoliaceae
Common Name: sweetbay
Part: bark
Class: 1

### *Mahonia aquifolium* (Pursh) Nutt. — Berberidaceae
Synonym: *Berberis aquifolium* Pursh.
### *Mahonia nervosa* (Pursh) Nutt.
### *Mahonia repens* (Lindl.) G. Don.
Common Name: Oregon grape
Other Common Names: Oregon grapeholly, Oregon barberry
Part: root
Class: 2b [3]
Notice: Berberine [17, 18] *See page 136.*
Ed. Note: Canadian regulations require bilingual label warning against use during pregnancy [3, (Welsh, 1995)] for M. *aquifolium* and M. *nervosa* and lists M. *aquifolium* and M. *nervosa* as unacceptable non-medicinal ingredients in oral use products (Michols, 1995).

### *Malva sylvestris* L. — Malvaceae
Common Name: high mallow
Other Common Names: malva
Part: leaf, flower
Class: 1

### *Mandragora officinarum* L. — Solanaceae
Common Name: mandrake
Part: root
Class: 3 [1, 3, 18, 22, 24, 29]
Notice: Atropine (0.4% total tropane alkaloids) [18, (Bruneton, 1995)] *See page 133.*
Ed. Note: Canadian regulations do not allow mandrake in foods [3].

---

Class 3: Herbs for which significant data exist to recommend the following labeling: "To be used only under the supervision of an expert qualified in the appropriate use of this substance." Labeling must include proper use information: dosage, contraindications, potential adverse effects and drug interactions, and any other relevant information related to the safe use of this substance.
Class 4: Herbs for which insufficient data are available for classification.

## *Maranta arundinacea* L.     Marantaceae
**Common Name:** arrowroot
**Part:** root
**Class:** 1
**Ed. Note:** Canadian regulations do not allow arrowroot as a non-medicinal ingredient for oral use products (Michols, 1995).

## *Marrubium vulgare* L.     Lamiaceae
**Common Name:** horehound
**Other Common Names:** white horehound
**Part:** herb
**Class:** 2b [18]
**Notice:** Emmenagogue/Uterine Stimulant [18] *See page 169.*
**Ed. Note:** Although recorded by Watt & Breyer-Brandwijk [25] as a purgative, an examination of their original references, cited as USP 24th and 25th ed., (Osol and Farrar, 1947; 1955) reveals only that "large doses are laxative." This is consistent with information presented by Chadha [6].

## *Matricaria recutita* L.     Asteraceae
**Common Name:** German chamomile
**Other Common Names:** Hungarian chamomile, true chamomile
**Part:** flower
**Class:** 1
**Ed. Note:** Minor side effects are recorded by several references. Highly concentrated hot tea is noted as emetic [6]; daily use is advised against by Van Hellemont [24].

The German Standard License, as presented by Wichtl [27], requires the following label warning: "The infusion should not be used near the eyes."

## *Medicago sativa* L.     Fabaceae
**Common Name:** alfalfa
**Part:** herb
**Class:** 1

## *Melia azedarach* L.     Meliaceae
**Common Name:** Chinatree, *ku lian pi*
**Part:** bark, root bark
**Class:** 3 [2, 10, 15, 16, 25]
**Notice:** Emetic [10, 25] *See page 167.*

---

Class 1: Herbs that can be safely consumed when used appropriately.
Class 2: Herbs for which the following use restrictions apply, unless otherwise directed by an expert qualified in the use of the described substance:
    (2a) For external use only;      (2b) Not to be used during pregnancy;
    (2c) Not to be used while nursing;      (2d) Other specific use restrictions as noted.

### *Melia toosendan* Sieb. & Zucc. — Meliaceae
**Common Name:** Sichuan pagoda tree, *chuan liang zi* (fruit); *ku lian pi* (root bark)
**Part:** root bark
**Class:** 3 [2, 14]

**Part:** fruit
**Class:** 1
**Ed. Note:** Significant toxicity of the fruit of the related species M. *azedarach* has been reported [2, 25].

### *Melissa officinalis* L. — Lamiaceae
**Common Name:** lemon balm
**Other Common Names:** balm
**Part:** leaf
**Class:** 1

### *Mentha x piperita* L. — Lamiaceae
**Common Name:** peppermint
**Part:** leaf
**Class:** 1

### *Mentha pulegium* L. — Lamiaceae
**Common Name:** European pennyroyal
**Other Common Names:** pennyroyal
**Part:** herb
**Class:** 2b [1, 6, 8, 18, 21, 22, 28]
**Notice:** Emmenagogue/Uterine Stimulant [6] *See page 169*.
**Ed. Note:** The leaf contains 1.0-2.0% essential oil consisting of 80.0-94.0% pulegone [18], a severe liver toxin when administered to rats in high dose [8, 18].

The notoriety of pennyroyal is due to the misuse of the essential oil. Death has been reported after consumption of as little as 15.0 ml (½ ounce) of the oil, and, in one instance, following consumption of an alcohol extract of pennyroyal herb over a two week period (Anderson et al., 1996). The last case involved a woman who was determined postmortem to have had an ectopic pregnancy, and toxic pulegone metabolites were reported to be present in the liver (Anderson et al., 1996).

M. *pulegium* and *Hedeoma pulegoides* have been historically interchangeable as the source for pennyroyal oil (Furia & Belanca, 1971).

---

Class 3: Herbs for which significant data exist to recommend the following labeling: "To be used only under the supervision of an expert qualified in the appropriate use of this substance." Labeling must include proper use information: dosage, contraindications, potential adverse effects and drug interactions, and any other relevant information related to the safe use of this substance.
Class 4: Herbs for which insufficient data are available for classification.

***Mentha spicata* L.**     **Lamiaceae**
**Common Name:** spearmint
**Part:** leaf
**Class:** 1


***Menyanthes trifoliata* L.**     **Menyanthaceae**
**Common Name:** bogbean
**Other Common Names:** buckbean
**Part:** leaf
**Class:** 2d - Contraindicated with diarrhea, dysentery, and colitis [5].
**Ed. Note:** Gastrointestinal distress is sometimes encountered, especially in high doses [24, 26].
    Regulated in the U.S. as an allowable flavoring agent in alcoholic beverages only [20].


***Microcos nervosa* (Lour.) S. Y. Hu.**     **Tiliaceae**
**Common Name:** microcos leaves, *bu za ye*
**Part:** leaf
**Class:** 4


***Mitchella repens* L.**     **Rubiaceae**
**Common Name:** squawvine
**Other Common Names:** partridge berry
**Part:** herb
**Class:** 1


***Monarda clinopodia* L.**     **Lamiaceae**
***Monarda didyma* L.**
***Monarda fistulosa* L.**
***Monarda pectinata* Nutt.**
***Monarda punctata* L.**
**Common Name:** beebalm (*M. clinopodia, M. didyma, M. fistulosa*), Oswego tea (*M. didyma*), wild bergamot (*M. clinopodia, M. fistulosa*), horsemint (*M. punctata*), spotted beebalm (*M. pectinata*)
**Part:** herb
**Class:** 2b [18, 28]
**Notice:** Emmenagogue/Uterine Stimulant [18, 28] *See page 169.*

---

Class 1: Herbs that can be safely consumed when used appropriately.
Class 2: Herbs for which the following use restrictions apply, unless otherwise directed by an expert qualified in the use of the described substance:
    (2a) For external use only;    (2b) Not to be used during pregnancy;
    (2c) Not to be used while nursing;   (2d) Other specific use restrictions as noted.

## *Morinda officinalis* How. — Rubiaceae
**Common Name:** morinda, *ba ji tian*
**Part:** root
**Class:** 1
**Ed. Note:** Bensky & Gamble [2] contraindicate morinda in difficult urination.

## *Morus alba* L. — Moraceae
**Common Name:** white mulberry, *sang bai pi* (root bark), *sang zhi* (twig), *sang shen* (fruit), *sang ye* (leaf)
**Part:** root bark, twig, fruit, leaf
**Class:** 1

## *Myrcia sphaerocarpa* DC. — Myrtaceae
**Common Name:** pedra hume caa
**Part:** herb
**Class:** 1

## *Myrica cerifera* L. — Myricaceae
## *Myrica pensylvanica* Lois.
**Common Name:** bayberry
**Other Common Names:** bayberry (*M. cerifica*, *M. pensylvanica*), wax myrtle (*M. cerifica*)
**Part:** bark, root
**Class:** 1
**Notice:** Tannins [18] *See page 155*.
**Ed. Note:** Canadian regulations do not allow bayberry as a non-medicinal ingredient for oral use products (Michols, 1995).

## *Myristica fragrans* Houtt. — Myristicaceae
**Common Name:** seed: nutmeg; aril: mace
**Part:** seed, aril
**Class:** 2b [4]; 3 [4, 6, 10, 12, 17, 18, 21, 22, 24, 26, 28, 29]
**Notice:** MAO Interaction (Truitt et al., 1963) *See page 173*.
Safrole (0.04-0.5%) [17, 18] *See page 152*.
**Ed. Note:** The classifications and concerns for this herb are based upon therapeutic use and may not be relevant to its consumption as a spice [4].

More than 5.0 grams of powdered nutmeg or mace affects the central nervous system, producing hallucinations, headache, dizziness, drowsiness, nausea, stomach pain, liver pain, excessive thirst, rapid pulse, anxiety, double vision, and sometime acute panic, coma, or death [17, 18, 21].

---

Class 3: Herbs for which significant data exist to recommend the following labeling: "To be used only under the supervision of an expert qualified in the appropriate use of this substance." Labeling must include proper use information: dosage, contraindications, potential adverse effects and drug interactions, and any other relevant information related to the safe use of this substance.
Class 4: Herbs for which insufficient data are available for classification.

### *Myroxylon balsamum* (L.) Harms var. *balsamum* — Fabaceae
### *Myroxylon balsamum* (L.) Harms var. *pereirae* (Royle) Harms.
**Common Name:** tolu balsam (*M. balsamum* [L.] Harms var.), balsam of Peru (*M. balsamum* [L.] var. *pereirae* [Royle] Harms)
**Other Common Names:** opobalsam, tolu
**Part:** oleo resin
**Class:** 2d - Contraindicated in febrile states [10]; may cause kidney irritation [10].
**Ed. Note:** Commission E [4] classifies *M. balsamum* for its external use only, limiting use to one week.

### *Nardostachys jatamansi* (D. Don) DC. — Valerianaceae
**Common Name:** jatamansi
**Other Common Names:** nard, spikenard
**Part:** rhizome, root
**Class:** 2b [6]
**Notice:** Emmenagogue/Uterine Stimulant [6] *See page 169*.

### *Nasturtium officinale* R.Br. — Brassicaceae
**Common Name:** watercress
**Part:** leaf
**Class:** 2b [18, 25]; 2d - Contraindicated in cases of gastric and duodenal ulcers, inflammatory kidney disorders, and for children under 4 years of age [4].
**Notice:** Emmenagogue/Uterine Stimulant [18, 25] *See page 169*.
**Ed. Note:** The classifications and concerns for this herb are based upon therapeutic use and may not be relevant to its consumption as a spice.
 Irritation of the gastric mucosa may develop if taken in large amounts or for an extended period of time [27].

### *Nelumbo nucifera* Gaertn. — Nymphaeaceae
**Common Name:** lotus, *lian zi* (seed), *lian xu* (stamen), *lian fang* (receptacle), *ou jie* (rhizome node), *he ye* (leaf), *lian zi xin* (embryo)
**Part:** seed
**Class:** 2d - Contraindicated in constipation and stomach distention [2, 29].
**Part:** stamen, receptacle, rhizome node, leaf, embryo
**Class:** 1

---

Class 1: Herbs that can be safely consumed when used appropriately.
Class 2: Herbs for which the following use restrictions apply, unless otherwise directed by an expert qualified in the use of the described substance:
  (2a) For external use only;   (2b) Not to be used during pregnancy;
  (2c) Not to be used while nursing;   (2d) Other specific use restrictions as noted.

## *Nepeta cataria* L. — Lamiaceae
Common Name: catnip
Other Common Names: catmint
Part: herb
Class: 2b [10]
Notice: Emmenagogue/Uterine Stimulant [10] *See page 169.*


## *Nereocystis luetkeana* (Mert.) Postels & Rupr. — Lessoniaceae
Common Name: kelp
Part: thallus
Class: 2d - Therapeutic use is not recommended in hyperthyroidism; long-term therapeutic use is not recommended [30]; take with adequate liquid [20].
Ed. Note: Specific labeling is required in the U.S. for all OTC drug products containing kelp [21] *See Appendix 2, page 165.*
Notice: Iodine *See page 145.*


## *Notopterygium incisum* Ting ex H. T. Chang. — Apiaceae
Common Name: notopterygium, *qiang huo*
Part: rhizome
Class: 1


## *Ocimum basilicum* L. — Lamiaceae
Common Name: basil
Other Common Names: sweet basil
Part: leaf
Class: 2b [4] ; 2c [4] ; 2d- not recommended for infants or toddlers, or for extended periods of time [4,17,27].
Notice: Estragole (70-85% of essential oil) *See page 143.*
Safrole (minor constituent) [17] *See page 152.*
Ed. Note: The classifications and concerns for this herb are based upon therapeutic use and may not be relevant to its consumption as a spice.


## *Oenothera biennis* L. — Onagraceae
Common Name: evening primrose
Part: herb, seed oil
Class: 1

---

Class 3: Herbs for which significant data exist to recommend the following labeling: "To be used only under the supervision of an expert qualified in the appropriate use of this substance." Labeling must include proper use information: dosage, contraindications, potential adverse effects and drug interactions, and any other relevant information related to the safe use of this substance.
Class 4: Herbs for which insufficient data are available for classification.

***Ophiopogon japonicus* (L. fil.) Ker-Gawl.**   Liliaceae
Common Name: dwarf-lilyturf root, *mai dong*
Part: root
Class: 1

***Oplopanax horridus* (Sm.) Miq.**   Araliaceae
Common Name: devil's club
Part: root
Class: 1

***Origanum majorana* L.**   Lamiaceae
Common Name: sweet marjoram
Part: leaf
Class: 1

***Origanum vulgare* L. subsp. *hirtum* (Link) Ietswaart.** Lamiaceae
Common Name: oregano
Part: leaf
Class: 1

***Paeonia lactiflora* Pall.**   Paeoniaceae
Common Name: white peony, *bai shao*
Other Common Names: common garden peony, peony root
Part: root
Class: 1

***Paeonia officinalis* L.**   Paeoniaceae
Common Name: European peony
Other Common Names: peony, piney, peony root
Part: root
Class: 1
Ed. Note: Canadian regulations do not allow European peony as a non-medicinal ingredient for oral use products (Michols, 1995).

***Paeonia suffruticosa* Andr.**   Paeoniaceae
Common Name: tree peony bark, *mu dan pi*
Other Common Names: tree peony, peony root

---

Class 1: Herbs that can be safely consumed when used appropriately.
Class 2: Herbs for which the following use restrictions apply, unless otherwise directed by an expert qualified in the use of the described substance:
    (2a) For external use only;    (2b) Not to be used during pregnancy;
    (2c) Not to be used while nursing;    (2d) Other specific use restrictions as noted.

Part: bark
Class: 2b [2, 29]

## *Panax ginseng* C. A. Mey.          Araliaceae
**Common Name:** oriental ginseng, *ren shen*
**Other Common Names:** Chinese ginseng, Korean ginseng, ginseng root
**Part:** root
**Class:** 2d - Contraindicated for hypertension [5].
**Standard Dose:** 0.6-3.0 grams, 1-3 times daily, eaten or prepared as tea [4, 5, 9, 27]; duration of use ranging from 3 weeks to 3 months is suggested at the therapeutic dosage range [9, 27].
**Ed. Note:** The use of red ginseng (processed by steaming) may potentiate the effects of caffeine and other stimulants [5]. Unprocessed ginseng is not considered as stimulating (Hobbs, 1996a).

Bradley [5] contraindicates ginseng during pregnancy. This caution is not consistent with traditional use in Asia, and data to support it are lacking [2, (Chin, 1991)].

Chang & But [7] report "Mild irritability and excitation were observed in persons who took 100 ml of 3.0% ginseng tincture; 200 ml of the tincture or large doses of ginseng powder could result in intoxication, giving rise to rose spots, pruritus, headache, vertigo, hyperpyrexia and bleeding, the last symptom being the characteristic manifestation of acute intoxication induced by ginseng. Three newborns were intoxicated after intake of 0.3-0.6 g of ginseng decocted; one died." They advise that "prolonged oral use of more than 0.3 g of ginseng powder might lead to insomnia, depression, headache, palpitation, hypertension, diminished sexual function and weight loss."

The statements from Chang & But are all translated from uncritical reports from China. The authors do not indicate the type of ginseng used, whether processed (red) or unprocessed (white). It is also known that prepared Chinese medicines can be adulterated with sulfites and pharmaceutical drugs. Adulteration and idiosyncratic reactions cannot be ruled out based on available literature [30].

Wichtl [27] reports "relatively rare" side effects, "...only with high doses and/or use over very long periods of time." These include "sleeplessness, nervousness, diarrhea (particularly in the morning), menopausal bleeding, and hypertony." Reynolds [21] presents a litany of side effects attributed to Siegel, whose work has been refuted due to methodological flaws [8, (Siegel, 1979), (Blumenthal, 1991)]. Three incidents of apparent hormonal-like effects in elderly women are recorded [12].

---

Class 3: Herbs for which significant data exist to recommend the following labeling: "To be used only under the supervision of an expert qualified in the appropriate use of this substance." Labeling must include proper use information: dosage, contraindications, potential adverse effects and drug interactions, and any other relevant information related to the safe use of this substance.
Class 4: Herbs for which insufficient data are available for classification.

### *Panax notoginseng* (Burk.) F.H.Chen. — Araliaceae
Common Name: *san qi* ginseng
Other Common Names: tienchi, tienchi ginseng, sanchi ginseng
Part: root
Class: 2b [2, 29]
Ed Note: *Panax notoginseng* is known to be adulterated with at least one other plant with a similar *pin yin* name [30].

### *Panax quinquefolius* L. — Araliaceae
Common Name: American ginseng, *ren shen*
Part: root
Class: 1

### *Papaver somniferum* L. — Papaveraceae
Common Name: poppyseed
Other Common Names: opium poppy
Part: seed
Class: 1
Ed. Note: The dried latex of the capsules is refined to produce opium, a controlled substance in many countries.

### *Parietaria judaica* L. — Urticaceae
### *Parietaria officinalis* L.
Common Name: pellitory of the wall
Part: herb
Class: 1

### *Parthenium integrifolium* L. — Asteraceae
Common Name: Missouri snakeroot
Other Common Names: prairie dock
Part: root
Class: 1

### *Passiflora incarnata* L. — Passifloraceae
Common Name: passion flower
Other Common Names: wild passion flower, maypop
Part: herb
Class: 1

---

Class 1: Herbs that can be safely consumed when used appropriately.
Class 2: Herbs for which the following use restrictions apply, unless otherwise directed by an expert qualified in the use of the described substance:
- (2a) For external use only;
- (2b) Not to be used during pregnancy;
- (2c) Not to be used while nursing;
- (2d) Other specific use restrictions as noted.

*Passiflora laurifolia* L.  Passifloraceae
**Common Name:** yellow granadilla
**Other Common Names:** passion flower herb
**Part:** herb
**Class:** 4


*Paullinia cupana* Humb., Bonpl. & Kunth.  Sapindaceae
**Common Name:** guarana
**Part:** seed
**Class:** 2d - Not recommended for excessive or long-term use [30].
**Notice:** Nervous System Stimulant (2.6-7.0% caffeine) [17, 18] *See page 174.*
Tannins (12.0%) [17] *See page 155.*
**Ed. Note:** The caffeine content of *P. cupana* is twice the amount found in coffee.


*Pausinystalia yohimbe* Pierre ex Beille.  Rubiaceae
**Synonym:** *Corynanthe yohimbi* Schum.
**Common Name:** yohimbe
**Other Common Names:** johimbe
**Part:** bark
**Class:** 2d - Contraindicated in existing liver and kidney diseases and in chronic inflammation of the sexual organs or prostate gland [4, 21, 22]; not recommended for excessive or long-term use; may potentiate pharmaceutical MAO-inhibitors [30].
**Standard Dose:** 5.0-6.0 mg of the contained alkaloid (yohimbine), three to four times daily [21, (Osol & Farrar, 1955)].
**Notice:** MAO Interaction [17, (Tyler, 1993)] *See page 173.*
Nervous System Stimulant [4, 17, 22] *See page 174.*
**Ed. Note:** In especially high dosages yohimbe can lower blood pressure and produce unpleasant digestive and central nervous system symptoms and may potentiate hypotensive drugs [4, 22, (Bruneton, 1995), (Osol & Farrar, 1955)].
   Canadian regulations do not allow yohimbe in foods [3].


*Pelargonium graveolens* L'Her. ex Aiton.  Geraniaceae
**Common Name:** rose geranium
**Part:** leaf
**Class:** 1

---

Class 3: Herbs for which significant data exist to recommend the following labeling: "To be used only under the supervision of an expert qualified in the appropriate use of this substance." Labeling must include proper use information: dosage, contraindications, potential adverse effects and drug interactions, and any other relevant information related to the safe use of this substance.
Class 4: Herbs for which insufficient data are available for classification.

### *Petroselinum crispum* (Mill.) Nym. ex A. W. Hill.  Apiaceae
**Common Name:** parsley
**Part:** leaf, root
**Class:** 2b [4, 5, 6, 25]; 2d - Contraindicated in inflammatory kidney disease[4, 5].
**Notice:** Emmenagogue/Uterine Stimulant [6, 25] *See page 169.*
**Ed. Note:** The classifications and concerns for this herb are based upon therapeutic use and may not be relevant to its consumption as a spice.

### *Peumus boldus* Molina.  Monimiaceae
**Common Name:** boldo
**Part:** leaf
**Class:** 2d - Contraindicated in serious liver conditions and obstruction of the bile duct; in persons with gallstones, the use of boldo should be under the supervision of a qualified practitioner [4].
**Ed. Note:** The use of boldo for liver ailments is widespread in South America for bile problems, gallstones, and liver conditions (Dragendorff, 1898).

Regulated in the U.S. as an allowable flavoring agent in alcoholic beverages only [20].

### *Pfaffia paniculata* (Mart.) Kuntze.  Amaranthaceae
**Common Name:** pfaffia
**Other Common Names:** suma
**Part:** root
**Class:** 1

### *Phellodendron amurense* Rupr.  Rutaceae
**Common Name:** phellodendron bark, *huang bai*
**Other Common Names:** amur cork tree
**Part:** bark
**Class:** 2b [30]
**Notice:** Berberine (0.6-2.5%) [2, 8, 17] *See page 136.*
**Ed. Note** - One case of a rash associated with ingestion of *huang bai* has been recorded [2].

### *Phellodendron chinense* Schneid.  Rutaceae
**Common Name:** phellodendron bark, *chuan huang bai*
**Other Common Names:** Chinese corktree, *huangpishu*
**Part:** bark
**Class:** 2b [30]
**Notice:** Berberine (4.0-8.0%) [2, 8, 17] *See page 136.*

---

Class 1: Herbs that can be safely consumed when used appropriately.
Class 2: Herbs for which the following use restrictions apply, unless otherwise directed by an expert qualified in the use of the described substance:
   (2a) For external use only;   (2b) Not to be used during pregnancy;
   (2c) Not to be used while nursing;   (2d) Other specific use restrictions as noted.

*Phoradendron leucarpum*                                 Viscaceae
  (Raf.) Reveal & M. C. Johnston
Common Name: American mistletoe
Part: herb
Class: 3 [16, 19]


*Phyllanthus emblica* L.                              Euphorbiaceae
Common Name: emblic
Other Common Names: ambal
Part: fruit
Class: 1


*Phytolacca americana* L.                           Phytolaccaceae
Common Name: poke
Part: root
Class: 3 [1, 9, 11, 15, 16, 25]
Notice: Lectins [9] *See page 146.*
Ed. Note: Canadian regulations do not allow poke in foods [3].


*Picrasma excelsa* (Swartz) Planch.               Simaroubaceae
Common Name: quassia
Other Common Names: Jamaican quassia
Part: bark, wood, root
Class: 2b [5, 18, 27]
Standard Dose: An infusion (tea) of 0.5 grams (±¼ tsp) of the bark 30 minutes prior to meals [27]. Decoction of 1-2 grams a day (Merck & Co., 1930).
Ed. Note: Consumption of large amounts of quassia can irritate the mucous membrane of the stomach and lead to vomiting [17, 18, 27].


*Pilocarpus jaborandi* Holmes.                       Rutaceae
*Pilocarpus microphyllus* Stapf.
*Pilocarpus pennatifolius* Lem.
Common Name: jaborandi
Other Common Names: Pernambuco jaborandi (*P. jaborandi*), Maranhao jaborandi (*P. microphyllus*), Paraguay jaborandi (*P. pennatifolius*)
Part: leaf
Class: 2b [30]; 3 [1, 22]

---

Class 3: Herbs for which significant data exist to recommend the following labeling: "To be used only under the supervision of an expert qualified in the appropriate use of this substance." Labeling must include proper use information: dosage, contraindications, potential adverse effects and drug interactions, and any other relevant information related to the safe use of this substance.
Class 4: Herbs for which insufficient data are available for classification.

*Pimenta dioica* (L.) Merr.                                  Myrtaceae
Common Name: allspice
Part: unripe fruit
Class: 1

*Pimpinella anisum* L.                                       Apiaceae
Common Name: anise
Part: fruit (commonly know as "seed")
Class: 2b [18]

*Pinellia ternata* (Thunb.) Briet.                           Araceae
Common Name: pinellia, *ban xia*
Part: prepared rhizome
Class: 2b [29, 30]; 2d - Contraindicated in all cases of bleeding [2] or with blood disorders [29].
Ed. Note: Although Yeung [29] contraindicates *Pinellia* in pregnancy, the revision of *Handbook of Chinese Herbs* does not restate this concern (Yeung, 1996).

*Pinus strobus* L.                                           Pinaceae
Common Name: white pine
Part: bark
Class: 1
Ed. Note: Regulated in the U.S. as an allowable flavoring agent in alcoholic beverages only [20].

*Piper cubeba* L. fil.                                       Piperaceae
Common Name: cubeb
Part: unripe fruit
Class: 2d - Contraindicated in nephritis [30].

*Piper methysticum* G. Forster.                              Piperaceae
Common Name: kava
Other Common Names: kava-kava, kava pepper
Part: root, rhizome
Class: 2b [4]; 2c [4]; 2d- Do not exceed recommended dose [30].

---

Class 1: Herbs that can be safely consumed when used appropriately.
Class 2: Herbs for which the following use restrictions apply, unless otherwise directed by an expert qualified in the use of the described substance:
    (2a) For external use only;    (2b) Not to be used during pregnancy;
    (2c) Not to be used while nursing;    (2d) Other specific use restrictions as noted.

**Standard Dose:** 2.0-4.0 grams as a decoction, up to three times daily [(Mitchell et al., 1983)]; 60-600 mg kavalactones (kavapyrones) per day [4, (Dentali, 1997)].
**Ed. Note:** Commission E [4] contraindicates kava for "endogenous depression". Some authors report that caution is required when driving or when operating other equipment [4, 26], and that simultaneous consumption with alcohol or barbiturates may potentiate inebriation [4, (Bruneton, 1995)].

Temporal limitations of from 4 weeks to 3 months are recommended in Australia and required in Germany; the following label has been recommended in Australia: "Warning: Do not exceed the stated dose" [1, 4]. Canadian regulations do not allow this herb as a non-medicinal ingredient for oral use products (Michols, 1995).

Excessive or extended consumption is reported to cause a scaly, yellowing skin condition which resolves when use is discontinued [4, (Bone, 1993/94), (Lewin, 1931)].

### *Piper nigrum* L.  Piperaceae
**Common Name:** black pepper (with cortex); white pepper (decorticated); green peppercorns (green with cortex)
**Other Common Names:** pepper, black peppercorns
**Part:** fruit
**Class:** 1

### *Plantago arenaria* Waldst. & Kit.  Plataginaceae
### *Plantago ovata* Forssk.
### *Plantago asiatica* L.
**Common Name:** psyllium (*P. asiatica*)
**Other Common Names:** French psyllium (*P. arenaria*), ispaghula (*P. ovata*)
**Part:** seed, seed husk
**Class:** 2d - Take with at least 250 ml (8 oz) of liquid [4, 5, 20, 27]; contraindicated in bowel obstruction [4, 5, 27]; take other drugs one hour prior to consumption of psyllium [4, 5].
**Standard Dose:** 2.5-10.0 grams (ca. ½-2 tsp), 2 to 3 times daily (Federal Register, 1985).
**Notice:** Bulk-forming Laxative [5, 17, 21, 27, 28] *See page 165.*
**Ed. Note:** Contemporary European references further contraindicate psyllium in "...diabetes mellitus which is difficult to regulate... ", noting that... "insulin-dependent diabetics may be required to decrease the insulin dosage" [4, 5].

Specific labeling is required in the U.S. for all OTC drug products containing psyllium [20] *See Appendix 2, page 165.*

---

Class 3: Herbs for which significant data exist to recommend the following labeling: "To be used only under the supervision of an expert qualified in the appropriate use of this substance." Labeling must include proper use information: dosage, contraindications, potential adverse effects and drug interactions, and any other relevant information related to the safe use of this substance.
Class 4: Herbs for which insufficient data are available for classification.

***Plantago lanceolata* L.**                                                            **Plataginaceae**
***Plantago major* L.**
***Plantago media* L.**
**Common Name:** plantain
**Other Common Names:** hoary plantain, English plantain
**Part:** leaf
**Class:** 1
**Ed. Note:** Instances of adulteration of *P. lanceolata* leaves with those of *Digitalis lanata* have been reported (Whitmore, 1997).

***Platycladus orientalis* (L.) Franco.**                       **Cupressaceae**
**Synonyms:** *Biota orientalis* (L.) Endl., *Thuja orientalis* L.
**Common Name:** oriental arborvitae, *bai zi ren* (seed), *ce bai ye* (cacumen)
**Other Common Names:** biota
**Part:** seed
**Class:** 1

**Part:** cacumen (leafy twigs)
**Class:** 2d - Not for long-term use; do not exceed recommended dose [30].
**Standard Dose:** 5.0-15.0 grams, raw or charred, daily as tea [2, (Yen, 1992)].
**Notice:** Thujone [2, 13, 29] *See page 158.*

***Platycodon grandiflorum* (Jacq.) A. DC.**               **Campanulaceae**
**Common Name:** balloon flower, *jie geng*
**Part:** root
**Class:** 2d - Contraindicated in cases of spitting blood and tuberculosis [2, 29]; administer "...cautiously to patients suffering from peptic ulcers" [14].
**Ed. Note:** Canadian regulations do not allow balloon flower as a non-medicinal ingredient for oral use products (Michols, 1995).

***Podophyllum hexandrum* Royle.**                           **Berberidaceae**
**Common Name:** Himalayan mayapple
**Other Common Names:** *Podophyllum emodi*
**Part:** root
**Class:** 2b [30]; 3 [1, 6, 9, 10, 12, 17, 21, 24, 28]
**Ed. Note:** Canadian regulations do not allow any part or derivative of this plant in foods [3].

---

Class 1: Herbs that can be safely consumed when used appropriately.
Class 2: Herbs for which the following use restrictions apply, unless otherwise directed by an expert qualified in the use of the described substance:
    (2a) For external use only;     (2b) Not to be used during pregnancy;
    (2c) Not to be used while nursing;     (2d) Other specific use restrictions as noted.

## *Podophyllum peltatum* L.  Berberidaceae
**Common Name:** mayapple
**Other Common Names:** American mandrake
**Part:** root
**Class:** 2b [30]; 3 [6, 9, 10, 11, 12, 17, 21, 24, 28]


## *Pogostemon cablin* (Blanco) Benth.  Lamiaceae
**Common Name:** patchouly, *huo xiang*
**Part:** herb
**Class:** 1


## *Polygala senega* L.  Polygalaceae
**Common Name:** Seneca snakeroot
**Other Common Names:** Senega snakeroot
**Part:** root
**Class:** 2b [10]; 2d - Contraindicated in gastritis and gastric ulcers [5]; not for long-term use [4, 9, 24].
**Standard Dose:** 0.5-1.0 grams three times daily [4, 5]; 1-3 grams a day (Merck & Co., 1930).
**Notice:** Emmenagogue/Uterine Stimulant [10] *See page 169*.
**Ed. Note:** Prolonged or therapeutic use may produce gastrointestinal irritation [4, 9, 24].


## *Polygala sibirica* L.  Polygalaceae
## *Polygala tenuifolia* Willd.
**Common Name:** polygala, *yuan zhi*
**Part:** root
**Class:** 2d - Contraindicated in ulcers and gastritis [2].


## *Polygonatum biflorum* (Walter.) Elliott.  Liliaceae
**Common Name:** Solomon's seal
**Part:** root
**Class:** 1


## *Polygonatum odoratum* (Mill.) Druce.  Liliaceae
**Common Name:** aromatic Solomon's seal, *yu zhu*
**Part:** rhizome
**Class:** 1

---

Class 3: Herbs for which significant data exist to recommend the following labeling: "To be used only under the supervision of an expert qualified in the appropriate use of this substance." Labeling must include proper use information: dosage, contraindications, potential adverse effects and drug interactions, and any other relevant information related to the safe use of this substance.
Class 4: Herbs for which insufficient data are available for classification.

*Polygonatum sibiricum* Red.                                             Liliaceae
**Common Name:** Siberian Solomon's seal, *huang jing*
**Part:** root
**Class:** 1

*Polygonum bistorta* L.                                                Polygonaceae
**Common Name:** bistort
**Part:** root
**Class:** 1

*Polygonum multiflorum* Thunb.                                Polygonaceae
**Common Name:** fo-ti, *he shou wu*
**Other Common Names:** Chinese knotweed, *ho shou wu*
**Part:** root
**Class:** 2d - Contraindicated with diarrhea [2, 29]; prepared root and stem may cause gastric distress [7]; raw root is cathartic [17].

*Populus balsamifera* L. var. *balsamifera.*                  Salicaceae
*Populus* x *jackii* Sarg.
**Common Name:** balm of Gilead (*P. balsamifera, P.* x *jackii*)
**Other Common Names:** poplar buds (*P. balsamifera*), tacamahac (*P. balsamifera*)
**Part:** resinous leaf buds
**Class:** 1
**Notice:** Salicylates [18] *See page 154.*
**Ed. Note:** Bradley [5] and Commission E [4] note hypersensitivity, which Bradley describes as "rare".

    Regulated in the U.S. as an allowable flavoring agent in alcoholic beverages only [20].

*Portulaca oleracea* L.                                            Portulacaceae
**Common Name:** purslane
**Part:** herb
**Class:** 2b [2, 14, 29]; **2d** - Individuals with a history of kidney stones should use this herb cautiously [30].
**Notice:** Oxalates (up to 1.7% oxalic acid) [6, 15, 17, 18, 25, (Duke, 1992)] *See page 148.*

---

Class 1: Herbs that can be safely consumed when used appropriately.
Class 2: Herbs for which the following use restrictions apply, unless otherwise directed by an expert qualified in the use of the described substance:
    (2a) For external use only;    (2b) Not to be used during pregnancy;
    (2c) Not to be used while nursing;    (2d) Other specific use restrictions as noted.

## *Potentilla erecta* (L.) Raeusch.   Rosaceae
**Common Name:** cinquefoil
**Other Common Names:** tormentil
**Part:** rhizome
**Class:** 1
**Notice:** Tannins (15.0-20.0%) [4, 24, 26, 27] *See page 155.*
**Ed. Note:** Sensitive individuals may experience gastric discomfort [4].

## *Primula veris* L.   Primulaceae
**Common Name:** cowslip
**Part:** flower, root
**Class:** 1
**Ed. Note:** Occasional gastric discomfort and nausea are recorded [4].

## *Prinsepia uniflora* Batalin.   Rosaceae
**Common Name:** prinsepia, *rui ren*
**Part:** seed
**Class:** 4

## *Prunella vulgaris* L.   Lamiaceae
**Common Name:** heal-all
**Other Common Names:** all-heal, self-heal
**Part:** herb
**Class:** 1

## *Prunus armeniaca* L.   Rosaceae
**Common Name:** apricot, *xing ren*
**Part:** seed
**Class:** 3 [2, 7, 11, 14, 16, 18, 22, 25, 29]
**Notice:** Cyanogenic Glycosides (amygdalin, up to 8.0%) [2, 14, 16, 18, 22, 25, 29] *See page 141.*
**Ed. Note:** Lethal dose is reported as from 7 to 10 kernels in children, 50 to 60 kernels in adults [2].

   *P. armeniaca* should not be sold in bulk; labels on all products should state: "Not for use by children."

---

Class 3: Herbs for which significant data exist to recommend the following labeling: "To be used only under the supervision of an expert qualified in the appropriate use of this substance." Labeling must include proper use information: dosage, contraindications, potential adverse effects and drug interactions, and any other relevant information related to the safe use of this substance.
Class 4: Herbs for which insufficient data are available for classification.

*Prunus dulcis* (Mill.) D.A. Webb var. *amara* (DC.)　　Rosaceae
H.E. Moore.
Common Name: bitter almond
Part: seed
Class: 3 [11, 16, 18, 30]
Notice: Cyanogenic Glycosides (amygdalin, 1.0-8.0%) [11, 16, 17, 18] *See page 141.*
Ed. Note: See *P. armeniaca*.

*Prunus mume* (Sieb.) Sieb. et Zucc.　　Rosaceae
Common Name: Japanese apricot, *wu mei*
Part: unripe fruit
Class: 1

*Prunus persica* (L.) Batsch.　　Rosaceae
Common Name: peach, *tao ren*
Part: seed
Class: 2b [2, 29]; 3 [6, 15, 16, 18, 25, 29]
Notice: Cyanogenic Glycosides (amygdalin, 2.0-6.0%) [6, 15, 16, 18, 19, 25, 29] *See page 141.*
Ed. Note: See *P. armeniaca*

Part: bark, leaf
Class: 3 [15, 18, 30]
Notice: Cyanogenic Glycosides (amygdalin, leaf: ca. 1.0%; bark: 2.0-3.0%) [15, 18, 28] *See page 141.*
Ed. Note: Although cyanogenic glycoside content of the leaf and bark is significantly less than in the seed, caution must still be observed, especially with children [15, 18].
　　Leaves regulated in the U.S. as an allowable flavoring agent in alcoholic beverages only, not to exceed 25.0 ppm prussic acid [20].

*Prunus serotina* Ehrh.　　Rosaceae
Common Name: wild cherry
Other Common Names: wild black cherry
Part: bark
Class: 2d - Not for long-term use; do not exceed recommended dose [30].
Standard Dose: 2.0-4.0 grams (Osol & Farrar, 1955).

---

Class 1: Herbs that can be safely consumed when used appropriately.
Class 2: Herbs for which the following use restrictions apply, unless otherwise directed by an expert qualified in the use of the described substance:
　　(2a) For external use only;　　(2b) Not to be used during pregnancy;
　　(2c) Not to be used while nursing;　(2d) Other specific use restrictions as noted.

**Notice:** Cyanogenic Glycosides (prunasin, yielding up to 0.15% hydrocyanic acid) [17, 18, 19, 28] *See page 141.*
**Ed. Note:** Canadian regulations do not allow wild cherry as a non-medicinal ingredient for oral use products (Michols, 1995).

*Prunus spinosa* L.  Rosaceae
**Common Name:** sloe
**Part:** flower, fruit
**Class:** 2d - Not for long-term use; do not exceed recommended dose [30].
**Standard Dose:** 1.0-2.0 grams (1.0-2.0 teaspoons) prepared as a tea, up to two times daily [27].
**Notice:** Cyanogenic Glycosides [18] *See page 141.*
**Ed. Note:** Cyanogenic glycosides are present only in the fresh flowers and the seeds [18].

*Pterocarpus santalinus* L. fil.  Fabaceae
**Common Name:** red sandalwood
**Other Common Names:** red saunders, sandalwood Padauk
**Part:** heart wood
**Class:** 1
**Ed. Note:** Regulated in the U.S. as an allowable flavoring agent in alcoholic beverages only [20].

*Ptychopetalum olacoides* Benth.  Olacaceae
*Ptychopetalum uncinatum* Anselmino.
**Common Name:** muira puama
**Part:** wood, root
**Class:** 1

*Pueraria lobata* (Willd.) Ohwi.  Fabaceae
*Pueraria thomsonii* Benth.
**Common Name:** kudzu (*P. lobata*), ge gen
**Other Common Names:** mealy kudzu (*P. thomsonii*)
**Part:** root
**Class:** 1

---

Class 3: Herbs for which significant data exist to recommend the following labeling: "To be used only under the supervision of an expert qualified in the appropriate use of this substance." Labeling must include proper use information: dosage, contraindications, potential adverse effects and drug interactions, and any other relevant information related to the safe use of this substance.
Class 4: Herbs for which insufficient data are available for classification.

## *Pulmonaria officinalis* L. — Boraginaceae
**Common Name:** lungwort
**Part:** herb
**Class:** 1
**Ed. Note:** Although De Smet [8] cites the presence of pyrrolizidine alkaloids in the herb, both Wren [28] and Wichtl [27] present gas chromatographic investigations which failed to detect these compounds in any of the samples tested.

## *Punica granatum* L. — Punicaceae
**Common Name:** pomegranate, *shi liu gen pi* (root bark); *shi liu pi* (fruit husk)
**Part:** root, stem bark
**Class:** 3 [6, 18, 22, 24]
**Notice:** Tannins (up to 25.0%) [18, 24] *See page 155.*
**Ed. Note:** Contains the toxic alkaloid pelletierine [21].

It was recommended by the Food Additives and Contaminants Committee that pomegranate root be prohibited for use in foods as a flavoring agent [21].

**Part:** fruit husk
**Class:** 2d - Contraindicated with diarrhea [6, 14, 18, 28]; not to be taken with fats or oils "...when taken to kill parasites" [2].
**Notice:** Tannins (up to 28.0%) [18] *See page 155.*

## *Quassia amara* L. — Simaroubaceae
**Common Name:** quassia
**Other Common Names:** Surinam quassia
**Part:** bark, wood, root
**Class:** 2b [18, 27]
**Standard Dose:** 1-2 grams a day of the wood as a decoction; average dose, 0.5 grams, 2-3 times daily (Merck & Co., 1930).
**Ed. Note:** Consumption of large amounts of quassia can irritate the mucous membrane of the stomach and lead to vomiting [17, 18, 27].

## *Quercus alba* L. — Fagaceae
## *Quercus robur* L.
## *Quercus petraea* (Mattuschka) Liebl.
**Common Name:** oak
**Other Common Names:** white oak bark (*Q. alba*)
**Part:** bark

---

Class 1: Herbs that can be safely consumed when used appropriately.
Class 2: Herbs for which the following use restrictions apply, unless otherwise directed by an expert qualified in the use of the described substance:
　　(2a) For external use only;　　(2b) Not to be used during pregnancy;
　　(2c) Not to be used while nursing;　　(2d) Other specific use restrictions as noted.

**Class: 2d** - Contraindicated for external use with extensive skin surface damage [4]; full baths with a significant amount of the tea are contraindicated in the following conditions: weeping eczema and skin damage over a large area; febrile and infectious disorders; cardiac insufficiency stages III and IV (NYHA); hypertonia stage IV (WHO) [4].
**Notice:** Tannins (6.0-20.0%) [4, 18, 27] *See page 155.*

*Quillaja saponaria* Molina.  Rosaceae
**Common Name:** quillaja
**Other Common Names:** soap tree
**Part:** bark
**Class: 2d** - Do not exceed recommended dose [18].
**Standard Dose:** 200 mg prepared as a tea [27].
**Notice:** GI Irritant [9, 17, 18, 21, 24] *See page 172.*
Oxalates [27] *See page 148.*
Tannins (10.0-15.0%) [27] *See page 155.*
**Ed. Note:** Large amounts of *Quillaja* are reported to produce liver damage, diarrhea, respiratory failure, stomach pain, convulsions, and coma [17, 18, 21, 24]. De Smet [9], however, notes that no support from a primary reference exists to substantiate these concerns.

Adequate protections should be taken when milling, as powder is caustic to mucosa [9, 24].

*Rehmannia glutinosa* Steud.  Gesneriaceae
**Common Name:** rehmannia, *sheng di huang*
**Other Common Names:** Chinese foxglove
**Part:** raw root (*sheng di huang*)
**Class: 2d** - Contraindicated with diarrhea and lack of appetite [29].

**Part:** cooked root (*shu di huang*)
**Class: 2d** - Contraindicated with diarrhea and indigestion [29]. Bensky and Gamble [2] state that side effects "...are mild and include diarrhea, abdominal pain, dizziness, lack of energy, and palpitations" [2].

*Rhamnus catharticus* L.  Rhamnaceae
*Rhamnus frangula* L.
**Common Name:** purging buckthorn (*R. catharticus*), alder buckthorn (*R. frangula*)
**Other Common Names:** buckthorn bark

---

Class 3: Herbs for which significant data exist to recommend the following labeling: "To be used only under the supervision of an expert qualified in the appropriate use of this substance." Labeling must include proper use information: dosage, contraindications, potential adverse effects and drug interactions, and any other relevant information related to the safe use of this substance.
Class 4: Herbs for which insufficient data are available for classification.

**Part:** fruit (*R. catharticus*); bark (*R. frangula*)
**Class:** 2b [4, 5, 6, 22, 27]; 2c [4, 5, 27]; 2d - Contraindicated in intestinal obstruction, abdominal pain of unknown origin, or any inflammatory condition of the intestines (appendicitis, colitis, Crohn's disease, irritable bowel, etc.) [4, 5, 9, 27]; and in children less than 12 years of age [4, 5, 9]; not for long-term use in excess of 8-10 days [4, 5, 9, 17, 26, 27].
**Standard Dose:** 2.0 grams (a scant teaspoon) infused as a tea [27].
**Notice:** Stimulant Laxative [4, 5, 6, 9, 10, 12, 17, 21, 22, 24, 26, 27, 28] *See page 177.*
**Ed. Note:** The bark of *R. frangula* must be aged for one to two years prior to use to destroy an emetic principle [9, 26, 28].

AHPA recommends the following label for products containing this herb in sufficient quantity to warrant such labeling: Do not use this product if you have abdominal pain or diarrhea. Consult a health care provider prior to use if you are pregnant or nursing. Discontinue use in the event of diarrhea or watery stools. Do not exceed recommended dose. Not for long-term use.

## *Rhamnus purshiana* DC.  Rhamnaceae
**Common Name:** cascara sagrada
**Part:** bark
**Class:** 2b [4, 5, 27]; 2c [4, 5, 27]; 2d - Contraindicated in intestinal obstruction, abdominal pain of unknown origin, or any inflammatory condition of the intestines (appendicitis, colitis, Crohn's disease, irritable bowel, etc.) [4, 5, 9, 21, 27]; and in children less than 12 years of age [4, 5, 9]; not for long-term use in excess of 8-10 days [4, 5, 6, 9, 17, 18, 27].
**Standard Dose:** "The individually correct dose is the smallest dosage necessary to maintain a soft stool" [4].

Adults: as a tea, steep 1.0-2.5 grams for 10 minutes and then strain [27, 28]; oral dose is 300-1000 mg in a single daily dose (Federal Register, 1985).

Children, ages 2-12: 150-500 mg in a single daily dose for children aged 2-12; consult a doctor for children under the age of 2 (Federal Register, 1985).
**Notice:** Stimulant Laxative [4, 5, 6, 9, 10, 12, 17, 18, 19, 21, 27, 28] *See page 177.*
**Ed. Note:** The bark of cascara sagrada must be aged for one year or heat treated prior to use [4, 9, 24, 27, 28].

AHPA recommends the following label for products containing this herb in sufficient quantity to warrant such labeling: Do not use this product if you have abdominal pain or diarrhea. Consult a health care provider prior to use if you are pregnant or nursing. Discontinue use in the event of diarrhea or watery stools. Do not exceed recommended dose. Not for long-term use.

---

Class 1: Herbs that can be safely consumed when used appropriately.
Class 2: Herbs for which the following use restrictions apply, unless otherwise directed by an expert qualified in the use of the described substance:
  (2a) For external use only;      (2b) Not to be used during pregnancy;
  (2c) Not to be used while nursing;   (2d) Other specific use restrictions as noted.

## Rheum officinale Baill.            Polygonaceae
## Rheum palmatum L.
## Rheum tanguticum Maxim. ex. Balf.

**Common Name:** Chinese rhubarb (*R. palmatum*), *da huang*
**Other Common Names:** rhubarb, turkey rhubarb (*R. officinale*, *R. palmatum*, *R. tanguticum*)
**Part:** rhizome and root
**Class:** 2b [2, 4, 5, 24, 27]; 2c [2, 4, 5, 24, 27]; 2d - Contraindicated in intestinal obstruction, abdominal pain of unknown origin, or any inflammatory condition of the intestines (appendicitis, colitis, Crohn's disease, irritable bowel, etc.) [4, 5, 9, 21, 22, 27]; and in children less than 12 years of age [4, 5, 9]; not for long-term use in excess of 8-10 days [4, 5, 9, 14, 17, 22, 24, 26, 27]; individuals with a history of kidney stones should use this herb cautiously [30].
**Standard Dose:** 1.0-5.0 grams daily, directly consumed or decocted as a tea [5], or up to 12.0 grams daily according to traditional Chinese authors [2, 13].
**Notice:** Oxalates (6.0% as calcium oxalate) [2, 4, 5, 9, 17, 24, 28, (Duke, 1992)] *See page 148.*
Stimulant Laxative [2, 4, 5, 9, 10, 13, 14, 17, 22, 24, 26, 28] *See page 177.*
Tannins (4.0-11.0%) [5, 13, 23, 28] *See page 155.*
**Ed. Note:** The petiole (leaf stem) of the related *R. rhaponticum*, which is commonly used for baking, is free of oxalates; the leaf blade must not be eaten [9, 12, 19].

AHPA recommends the following label for products containing this herb in sufficient quantity to warrant such labeling: Do not use this product if you have abdominal pain or diarrhea. Consult a health care provider prior to use if you are pregnant or nursing. Discontinue use in the event of diarrhea or watery stools. Do not exceed recommended dose. Not for long-term use.

## *Rhodymenia palmetta* (J. F. Lamour.) Grev.     Rhodymeniaceae

**Common Name:** dulse
**Part:** thallus
**Class:** 2d - Therapeutic use is not recommended in hyperthyroidism; long-term therapeutic use is not recommended [30].
**Notice:** Iodine (0.7% of dry weight) (Chapman, 1970) *See page 145.*

---

Class 3: Herbs for which significant data exist to recommend the following labeling: "To be used only under the supervision of an expert qualified in the appropriate use of this substance." Labeling must include proper use information: dosage, contraindications, potential adverse effects and drug interactions, and any other relevant information related to the safe use of this substance.
Class 4: Herbs for which insufficient data are available for classification.

*Rhus coriaria* L.            Anacardiaceae
*Rhus glabra* L.
**Common Name:** sumac
**Part:** fruit
**Class:** 1

**Part:** bark
**Class:** 1
**Standard Dose:** A teaspoon steeped as a tea, 1-2 times daily (Lust, 1974).
**Notice:** Tannins (24.0-35.0%) [18] *See page 155*.
**Ed. Note:** Excessive use of the bark will cause catharsis [10].

The following label has been recommended in Australia for *Rhus glabra*: "Do not exceed the stated dose" [1].

*Ribes nigrum* L.            Grossulariaceae
**Common Name:** black currant
**Other Common Names:** cassis
**Part:** fruit
**Class:** 1

*Ricinus communis* L.            Euphorbiaceae
**Common Name:** castor
**Other Common Names:** palma christi
**Part:** seed oil
**Class:** 2b [6, 21, 30]; 2d - Contraindicated in intestinal obstruction and abdominal pain of unknown origin [21]; not for long-term use in excess of 8-10 days [30].
**Standard Dose:**
    Adults: 5.0-20.0 ml per dose, not to exceed 60.0 ml per day.
    Children: 4.0-12.0 ml [17, 28, (Osol & Farrar, 1955)].
**Notice:** Lectins [18] *See page 146*.
Stimulant Laxative [6, 10, 13, 17, 18, 21] *See page 177*.
**Ed. Note:** Chadha [6] and Felter & Lloyd [10] assert that castor oil is "...especially adapted..." for pregnancy. However, other references state that castor oil must be used with caution during pregnancy and menstruation [21, (Osol & Farrar, 1955)].

Midwives sometimes use castor oil to induce labor, a practice also described in pharmaceutical texts [30, (Osol & Farrar, 1955)].

---

Class 1: Herbs that can be safely consumed when used appropriately.
Class 2: Herbs for which the following use restrictions apply, unless otherwise directed by an expert qualified in the use of the described substance:
    (2a) For external use only;     (2b) Not to be used during pregnancy;
    (2c) Not to be used while nursing;   (2d) Other specific use restrictions as noted.

In one of their few unreferenced statements, Leung & Foster [17] state that castor oil should not be given with potentially toxic oil-soluble anthelmintics. Only one additional reference raises this concern, but their source is given as Leung's first edition [28].

*Rosa alba* L.  Rosaceae
*Rosa canina* L.
*Rosa centifolia* L.
*Rosa damascena* L.
*Rosa gallica* L.
Common Name: white rose (R. *alba*), dog rose (R. *canina*), provence rose (R. *centifolia*), damask rose (R. *damascena*)
Part: fruit
Class: 1

*Rosa rugosa* L.  Rosaceae
Common Name: rugose rose, *mei gui hua*
Part: flower
Class: 1

*Rosmarinus officinalis* L.  Lamiaceae
Common Name: rosemary
Part: leaf
Class: 2b [6, 9, 24, 25, 27, 30]
Notice: Abortifacient [6, 25] *See page 163*.
Emmenagogue/Uterine Stimulant [6, 10] *See page 169*.
Ed. Note: The classifications and concerns for this herb are based upon therapeutic use and dosage and may not be relevant to its consumption as a spice.

*Rubus fruticosus* L.  Rosaceae
*Rubus idaeus* L.
*Rubus strigosus* Michx.
Common Name: blackberry (R. *fruticosus*), raspberry (R. *idaeus*, R. *strigosus*)
Other Common Names: red raspberry (R. *idaeus*)
Part: leaf, fruit
Class: 1

---

Class 3: Herbs for which significant data exist to recommend the following labeling: "To be used only under the supervision of an expert qualified in the appropriate use of this substance." Labeling must include proper use information: dosage, contraindications, potential adverse effects and drug interactions, and any other relevant information related to the safe use of this substance.
Class 4: Herbs for which insufficient data are available for classification.

*Rubus chingii* Hu.      Rosaceae
*Rubus suavissimus* S.Lee.
**Common Name:** sweet tea, *fu pen zi, tian cha*
**Part:** fruit
**Class:** 2d - Contraindicated in urinary difficulty [2, 29].


*Rumex acetosa* L.      Polygonaceae
*Rumex acetosella* L.
**Common Name:** sorrel, sheep sorrel
**Other Common Names:** field sorrel
**Part:** leaf
**Class:** 2d - Individuals with a history of kidney stones should use this herb cautiously [30].
**Notice:** Oxalates (0.3% oxalic acid in the leaf of *R. acetosa*) [15, 18, 28, (Duke, 1992)] *See page 148.*
Tannins (7.0-15.0%) [18] *See page 155.*


*Rumex crispus* L.      Polygonaceae
*Rumex obtusifolius* L.
**Common Name:** yellow dock, broad-leaved dock
**Other Common Names:** curled dock, curly dock
**Part:** root
**Class:** 2d - Individuals with a history of kidney stones should use this herb cautiously [30].
**Notice:** Oxalates [15, 18, 28] *See page 148.*
Tannins (12.0-20.0%) [18] *See page 155.*
**Ed. Note:** Although *Rumex* contains a small amount of anthraquinone glycosides (3.0-4.0%), it has, at most, a mild laxative effect [18, 28].


*Rumex hymenosepalus* Torr.      Polygonaceae
**Common Name:** canaigre
**Part:** root
**Class:** 1
**Notice:** Tannins (30.0-35.0%) [18] *See page 155.*


*Ruscus aculeatus* L.      Liliaceae
**Common Name:** butcher's broom, box holly
**Part:** root
**Class:** 1

---

Class 1: Herbs that can be safely consumed when used appropriately.
Class 2: Herbs for which the following use restrictions apply, unless otherwise directed by an expert qualified in the use of the described substance:
    (2a) For external use only;      (2b) Not to be used during pregnancy;
    (2c) Not to be used while nursing;      (2d) Other specific use restrictions as noted.

## *Ruta graveolens* L. Rutaceae
**Common Name:** rue
**Other Common Names:** herb-of-grace
**Part:** herb
**Class:** 2b [3, 6, 18, 25, 28, 30]; **2d** - Contraindicated in poor kidney function [24]; avoid prolonged exposure to sunlight [30].
**Notice:** Abortifacient [4, 6, 10, 17, 25, 26] *See page 163*.
Emmenagogue/Uterine Stimulant [6, 10, 17, 18, 25, 26, 28] *See page 169*.
GI Irritant [4, 10, 28] *See page 172*.
Photosensitizing [4, 24] *See page 176*.
**Ed. Note:** Current U.S. regulations allow the herb to be used in food only if the concentration is less than 2.0 ppm [20]. Canadian regulations do not allow it as a non-medicinal ingredient for oral use products (Michols, 1995).

## *Salix alba* L. Salicaceae
**Common Name:** white willow
**Part:** bark
**Class:** 1
**Notice:** Salicylates (1.5-11.0% total phenolic glycosides) [5, 18, 26, 27, 28] *See page 154*.
Tannins (8.0-20.0%) [18, 27] *See page 155*.
**Ed. Note:** Other species of *Salix* are used interchangeably in trade, with salicin content ranging from 1.5-11.0% [5, 27].

## *Salvia columbariae* Benth. Lamiaceae
## *Salvia hispanica* L.
**Common Name:** California chia
**Other Common Names:** chia
**Part:** seed
**Class:** 1

## *Salvia miltiorrhiza* Bunge. Lamiaceae
**Common Name:** red-rooted sage, *dan shen*
**Other Common Names:** salvia root
**Part:** root
**Class:** 1
**Ed. Note:** Administration of tinctures of this herb reportedly leads to pruritus, stomachache, or reduced appetite [2].

---

Class 3: Herbs for which significant data exist to recommend the following labeling: "To be used only under the supervision of an expert qualified in the appropriate use of this substance." Labeling must include proper use information: dosage, contraindications, potential adverse effects and drug interactions, and any other relevant information related to the safe use of this substance.
Class 4: Herbs for which insufficient data are available for classification.

## *Salvia officinalis* L.                                                         **Lamiaceae**

**Common Name:** sage
**Other Common Names:** Dalmatian sage, garden sage, common sage
**Part:** leaf
**Class:** 2b [27, 30]; 2d - Not for long-term use; do not exceed recommended dose [30].
**Standard Dose:** 4.0-6.0 grams daily [4].
**Notice:** Thujone (0.5-1.5%) [17, (Farrell, 1990)] *See page 158.*
**Ed. Note:** The classifications and concerns for this herb are based upon therapeutic use and dosage and may not be relevant to its consumption as a spice.

The alcohol extract of sage is contraindicated in pregnancy [4].

Sage oil contains more thujone than absinthium oil, yet it has not been reported to be toxic [17].

This herb has been used traditionally to reduce lactation [17].

## *Salvia sclarea* L.                                                              **Lamiaceae**

**Common Name:** clary sage
**Other Common Names:** clary, muscatel sage
**Part:** herb
**Class:** 1

## *Sambucus canadensis* L.                                            **Caprifoliaceae**

**Common Name:** American elder
**Other Common Names:** elderberry, common elderberry, elder flower
**Part:** flower, ripe fruit
**Class:** 1
**Ed. Note:** Unripe fruit contains cyanogenic glycosides which may cause vomiting [11].

## *Sambucus nigra* L.                                                         **Caprifoliaceae**

**Common Name:** European elder
**Other Common Names:** elder flower
**Part:** flower, ripe fruit
**Class:** 1
**Ed. Note:** The raw and unripe fruit, the seeds, the bark, and the leaves of *S. nigra* and related species *S. racemosa* contain the cyanogenic glycoside sambunigrin, ingestion of which may cause vomiting or severe diarrhea [16, 26, 27].

---

Class 1: Herbs that can be safely consumed when used appropriately.
Class 2: Herbs for which the following use restrictions apply, unless otherwise directed by an expert qualified in the use of the described substance:
    (2a) For external use only;     (2b) Not to be used during pregnancy;
    (2c) Not to be used while nursing;     (2d) Other specific use restrictions as noted.

## *Sanguinaria canadensis* L.          Papaveraceae
**Common Name:** bloodroot
**Other Common Names:** red puccoon, red root
**Part:** root
**Class:** 2b [30]; 2d - May cause nausea and vomiting [10, (Osol & Farrar, 1955)].
**Notice:** Berberine [8, 17, 18] *See page 136.*
Emetic [10, (Osol & Farrar, 1955)] *See page 167.*
GI Irritant [10, 30] *See page 172.*
**Ed. Note:** Powerful emesis is produced by as little as one gram [10, (Osol & Farrar, 1955)].

The presence of the toxic alkaloid sanguinarine suggests that the plant should not be used in large amounts [17, 18, 19, 21, 28].

Canadian regulations do not allow bloodroot in foods [3]. The following labeling has been proposed for products sold in Australia: "May affect glaucoma treatment. Do not exceed the stated dose" [1].

## *Sanicula europaea* L.          Apiaceae
**Common Name:** European sanicle
**Other Common Names:** sanicle
**Part:** herb
**Class:** 1

## *Santalum album* L.          Santalaceae
**Common Name:** sandalwood, *tan xiang*
**Other Common Names:** East Indian sandalwood, white sandalwood, white saunders, yellow saunders, yellow sandalwood
**Part:** wood
**Class:** 2d - Contraindicated in diseases involving the parenchyma of the kidney [4]; not for use beyond 6 weeks without consultation with a physician [4].

## *Sassafras albidum* (Nutt.) Nees          Lauraceae
**Common Name:** sassafras
**Other Common Names:** filé (leaf only)
**Part:** leaf, root
**Class:** 2d - Not for long-term use; do not exceed recommended dose [30].
**Standard Dose:** 10.0 grams of the root bark (Powers et al., 1942); 2.0-4.0 ml of an liquid extract of root bark [28].

---

Class 3: Herbs for which significant data exist to recommend the following labeling: "To be used only under the supervision of an expert qualified in the appropriate use of this substance." Labeling must include proper use information: dosage, contraindications, potential adverse effects and drug interactions, and any other relevant information related to the safe use of this substance.
Class 4: Herbs for which insufficient data are available for classification.

**Notice:** Safrole (5.0-8.0% in the root bark; less than 1.0% in the root) [17, 18, 28, (Cook & Martin, 1948)] *See page 152.*

**Ed. Note:** Both U.S. and Canadian regulations allow the use of *Sassafras* in foods only if safrole-free [3, 20].

### *Satureja hortensis* L.  Lamiaceae
### *Satureja montana* L.
**Common Name:** summer savory, winter savory
**Part:** leaf
**Class:** 1

**Ed. Note:** The classifications and concerns for this herb are based upon therapeutic use and may not be relevant to its consumption as a spice.

Van Hellemont [24] claims that internal use of *S. hortensis* (summer savory) may cause skin eruption. No consumption level was noted, and no primary data was included in this report.

### *Saussurea lappa* (Dcne.) C.B. Clarke.  Asteraceae
**Common Name:** costus, *mu xiang*
**Part:** root
**Class:** 1

### *Schinus terebinthifolia* Raddi.  Anacardiaceae
### *Schinus molle* L.
**Common Name:** pink peppercorns
**Other Common Names:** Brazil peppertree, pink pepper
**Part:** bark, fruit
**Class:** 1

**Notice:** GI Irritant [11, 16, 25] *See page 172.*

**Ed. Note:** Although both species are toxic in large amounts, the fruit is commonly used as a condiment [6, 25].

### *Schisandra chinensis* (Turcz.) Baill.  Schisandraceae
**Common Name:** schizandra, *wu wei zi*
**Other Common Names:** magnolia vine
**Part:** fruit
**Class:** 1

**Ed. Note:** Rare side effects of appetite suppression, stomach upset, and urticaria are recorded [17].

---

Class 1: Herbs that can be safely consumed when used appropriately.
Class 2: Herbs for which the following use restrictions apply, unless otherwise directed by an expert qualified in the use of the described substance:
    (2a) For external use only;    (2b) Not to be used during pregnancy;
    (2c) Not to be used while nursing;    (2d) Other specific use restrictions as noted.

*Scopolia carniolica* Jacq.  Solanaceae
**Common Name:** scopolia
**Other Common Names:** belladonna, Russian belladonna
**Part:** root, rhizome
**Class:** 3 [1, 4, 22]


*Scrophularia marilandica* L.  Scrophulariaceae
*Scrophularia nodosa* L.
**Common Name:** figwort
**Part:** herb, root
**Class:** 2d - Contraindicated in ventricular tachycardia (Mitchell et al., 1983).


*Scutellaria baicalensis* Georgi.  Lamiaceae
**Common Name:** Baikal skullcap, *huang qin*
**Other Common Names:** scute
**Part:** root
**Class:** 1


*Scutellaria lateriflora* L.  Lamiaceae
**Common Name:** skullcap
**Other Common Names:** scullcap
**Part:** herb
**Class:** 1
**Ed. Note:** Reports of toxicity are likely due to adulteration of skullcap with germander (*Teucrium* spp.), which has been reported to cause hepatotoxicity. De Smet [9] cites an "...increasing number of case reports to suggest that the ingestion of skullcap-containing preparations can induce hepatotoxic reactions." Yet in an appended clarification, he acknowledges the adulteration of skullcap with species of *Teucrium*. This adulteration, along with recent reports of hepatitis associated with *Teucrium* consumption, lead De Smet to conclude that it is "...unclear at the moment, whether the hepatotoxic effects that have been associated with preparations containing skullcap should be attributed to *Scutellaria*, *Teucrium*, or both."


*Selenicereus grandiflorus* (L.) Britton & Rose.  Cactaceae
**Common Name:** night-blooming cereus
**Part:** flower, stem
**Class:** 1

---

Class 3: Herbs for which significant data exist to recommend the following labeling: "To be used only under the supervision of an expert qualified in the appropriate use of this substance." Labeling must include proper use information: dosage, contraindications, potential adverse effects and drug interactions, and any other relevant information related to the safe use of this substance.
Class 4: Herbs for which insufficient data are available for classification.

**Ed. Note:** Although Wren [28] reports a digitalis-like effect for the alkaloid cactine, List & Hörhammer [18] claim it to be non-cumulative. The long history of use and the lack of reports of human toxicity suggest that such an effect is likely to be minimal.

## *Senna alexandrina* P. Mill.                                           Fabaceae
## *Senna obtusifolia* (L.) Irwin & Barneby.
**Synonym:** *Cassia obtusifolia* L.
## *Senna tora* (L.) Roxb.
**Synonym:** *Cassia tora* L.
**Common Name:** senna
**Other Common Names:** Indian senna
**Part:** leaf
**Class:** 2b [2, 4, 5, 24, 27]; 2c [2, 4, 5, 24]; 2d - Contraindicated in intestinal obstruction, abdominal pain of unknown origin, or any inflammatory condition of the intestines (appendicitis, colitis, Crohn's disease, irritable bowel, etc.) [4, 5, 9, 21, 27]; in hemorrhoids [24]; and in children less than 12 years of age [4, 9]; not for long-term use in excess of 8-10 days [4, 5, 9, 17, 18, 26, 27].
**Standard Dose:** Dried leaf: 0.5-3.0 grams infused in hot (not boiling) water for 10-15 minutes or in cold water for 10-12 hours [2, 13, 26, 27].
**Notice:** Stimulant Laxative [4, 5, 9, 17, 18, 21, 24, 26, 27, 28, 29] *See page 177.*
**Ed. Note:** The dianthrone glycosides responsible for the laxative action of *Senna* are formed during the process of dehydration, as they are not found in the fresh leaf [17, (Bruneton, 1995)].

AHPA recommends the following label for products containing this herb in sufficient quantity to warrant such labeling: Do not use this product if you have abdominal pain or diarrhea. Consult a health care provider prior to use if you are pregnant or nursing. Discontinue use in the event of diarrhea or watery stools. Do not exceed recommended dose. Not for long-term use.

**Part:** fruit
**Class:** 2b [2, 4, 5, 24, 27]; 2c [2, 5, 24]; 2d - Contraindicated in intestinal obstruction, abdominal pain of unknown origin, or any inflammatory condition of the intestines (appendicitis, colitis, Crohn's disease, irritable bowel, etc.) [4, 5, 9, 21, 27]; in hemorrhoids [24]; and in children less than 12 years of age [4, 9]; not for long-term use in excess of 8-10 days [4, 5, 9, 17, 18, 26, 27].
**Notice:** Stimulant Laxative [4, 5, 9, 17, 18, 21, 24, 26, 27, 28, 29] *See page 177.*

---

Class 1: Herbs that can be safely consumed when used appropriately.
Class 2: Herbs for which the following use restrictions apply, unless otherwise directed by an expert qualified in the use of the described substance:
    (2a) For external use only;     (2b) Not to be used during pregnancy;
    (2c) Not to be used while nursing;     (2d) Other specific use restrictions as noted.

**Ed. Note:** Most authorities [26, 28, 27] note that a preparation of the fruit (pods) acts more gently than the leaves. However, long-term concerns regarding the use of stimulant laxatives are still relevant.

There is a lack of unanimity in international labeling regulations for use of senna leaves and fruit by pregnant women and nursing mothers. In Britain as well as Germany, there are clearly stated contraindications for these conditions [5, 27]. No label restriction is required in the tentative final monograph for OTC use in the United States (Federal Register, 1985), and, in fact, the entire plant is listed in the Code of Federal Regulations as an acceptable flavoring agent [20]. Although senna is excluded from the French regulation against use of anthranoid laxatives during lactation, and the World Health Organization regards its use while breast feeding to be safe for the infant, some contemporary authors continue to prescribe caution [9].

## *Serenoa repens* (W. Bartram.) Small.   Arecaceae
**Common Name:** saw palmetto
**Other Common Names:** sabal
**Part:** fruit
**Class:** 1
**Ed. Note:** Rare cases of stomach problems are recorded [4].

Commission E [4] suggests regular consultation with a physician when using this herb for treatment of enlarged prostate. This caution is based on the stated assumption that *Serenoa* treats only symptoms without eliminating hypertrophic concern [4, 9].

## *Sesamum orientale* L.   Pedaliaceae
**Common Name:** sesame
**Part:** seed
**Class:** 1
**Ed. Note:** Anaphylaxis associated with sesame consumption has been described, especially in persons with a known history of asthma or with other anaphylactic sensitivity (James et al., 1991).

## *Silybum marianum* (L.) Gaertn.   Asteraceae
**Common Name:** milk thistle
**Part:** seed
**Class:** 1

---

Class 3: Herbs for which significant data exist to recommend the following labeling: "To be used only under the supervision of an expert qualified in the appropriate use of this substance." Labeling must include proper use information: dosage, contraindications, potential adverse effects and drug interactions, and any other relevant information related to the safe use of this substance.
Class 4: Herbs for which insufficient data are available for classification.

*Sinapis alba* L. **Brassicaceae**
Common Name: mustard
Other Common Names: yellow mustard
Part: seed
Class Internal Use: 1
Ed. Note: It is logical to assume that the contraindication with kidney disorders noted by Commission E [4] for external use may also apply to internal use. Ingestion of large quantities can cause irritant poisoning [25].

Class External Use: 2d - Use not to exceed 2 weeks [4]; not for external use with children under 6 years of age [4, 10, 18, 28]; contraindicated with kidney disorders [4].
Ed. Note: Severe burns can occur if applied for a prolonged period of time (over 15-30 minutes).

*Smilax febrifuga* Kunth. **Smilacaceae**
*Smilax medica* Schlechtend. & Cham.
*Smilax ornata* Lem.
*Smilax regelii* Killip & Morton.
Common Name: sarsaparilla
Part: root
Class: 1
Ed. Note: Commission E [4] warns that "taking sarsaparilla preparation leads to gastric irritation and temporary kidney impairment" and advises of potential drug interactions with hypnotics, digitalis glycosides, and bismuth. However, no other reference substantiates these concerns.

*Smilax glauca* Walter. **Smilacaceae**
Common Name: sawbrier
Part: root
Class: 1

*Solidago virgaurea* L. **Asteraceae**
*Solidago canadensis* L.
*Solidago gigantea* Ait.
Common Name: goldenrod
Part: herb
Class: 2d - In chronic kidney disorders, a practitioner should be consulted [27].

---

Class 1: Herbs that can be safely consumed when used appropriately.
Class 2: Herbs for which the following use restrictions apply, unless otherwise directed by an expert qualified in the use of the described substance:
    (2a) For external use only;    (2b) Not to be used during pregnancy;
    (2c) Not to be used while nursing;    (2d) Other specific use restrictions as noted.

## *Sophora flavescens* Ait.      Fabaceae
**Common Name:** shrubby sophora, *ku shen*
**Part:** root
**Class:** 1
**Ed. Note:** Occasional side effects such as mild dizziness, nausea, vomiting, and constipation may occur [13].


## *Spigelia marilandica* L.      Loganiaceae
**Common Name:** pinkroot
**Part:** root
**Class:** 2d - Not for long-term use. Do not exceed recommended dose [1, 10, 16, 22].
**Standard Dose:**
   Adults: 2.0-5.0 grams, morning and evening, given with a strong purgative, such as senna [28].
   Children over 4 years: 0.5-4.0 grams, morning and evening, given with a strong purgative, such as senna [28].
**Ed. Note:** "Various unpleasant symptoms" are reported when use is not accompanied with catharsis, and the potential toxicity of the fresh root has been especially emphasized [10, 22]. However, the *United States Dispensatory*, 25th ed. (Osol & Farrar, 1955) reported the drug was commonly used throughout the country as a vermifuge, and side effects were ..."almost unheard of."


## *Spilanthes acmella* L. ex J. A. Murray.      Asteraceae
## *Spilanthes oleracea* L.
**Common Name:** toothache plant
**Other Common Names:** para' cress, Paraguay cress
**Part:** herb
**Class:** 1


## *Stachys officinalis* (L.) Tervis.      Lamiaceae
**Common Name:** wood betony
**Part:** herb
**Class:** 1

---

Class 3: Herbs for which significant data exist to recommend the following labeling: "To be used only under the supervision of an expert qualified in the appropriate use of this substance." Labeling must include proper use information: dosage, contraindications, potential adverse effects and drug interactions, and any other relevant information related to the safe use of this substance.
Class 4: Herbs for which insufficient data are available for classification.

## *Stellaria media* (L.) Vill. — Caryophyllaceae
**Common Name:** chickweed
**Part:** herb
**Class:** 1

**Ed. Note:** There is one report of an alleged case of nitrate toxicity associated with chickweed which resulted in a mild form of paralysis [6]. The possibility that this event was due to environmental factors, such as harvesting from fields where synthetic fertilizers had been used [6, 30], should be examined.

## *Stephania tetrandra* S. Moore. — Menispermaceae
**Common Name:** stephania root, *fang ji*
**Other Common Names:** *han fang ji*
**Part:** root
**Class:** 1

**Ed. Note:** Cases of nephrotoxicity were reported from ingestion of diet products sold in Europe which listed *Stephania* as an ingredient. These were analyzed and found to contain aristolochic acid (Moffet, 1995), which is not a constituent of *S. tetrandra* [2, 13, 14]. The likely adulterant, *Aristolochia fangchi*, traded as *guang fang ji* [2], is considered interchangeable with *S. tetrandra* [13, 29], which is sold as *han fang ji* [2].

Canadian regulations do not allow stephania root as a non-medicinal ingredient for oral use products (Michols, 1995).

## *Stevia rebaudiana* (Bertoni) Hemsl. — Asteraceae
**Common Name:** stevia
**Other Common Names:** Paraguayan sweet herb, sweetleaf
**Part:** leaf
**Class:** 1

**Ed. Note:** The herb has been extensively used as a sweetener for beverages, originally in Paraguay and, in this century, in other South American countries and in Asia [17]. In the U.S., the FDA issued an import alert in May 1991 identifying stevia as an "unsafe food additive" (Blumenthal, 1992). The import alert was revised on September 18, 1995 to allow the entry of stevia "explicitly labeled as a dietary supplement or for use as a dietary ingredient of a dietary supplement" (Linsley, 1995).

Stevia is not currently allowed as a non-nutritional ingredient in Canada [3].

---

Class 1: Herbs that can be safely consumed when used appropriately.
Class 2: Herbs for which the following use restrictions apply, unless otherwise directed by an expert qualified in the use of the described substance:
 (2a) For external use only;  (2b) Not to be used during pregnancy;
 (2c) Not to be used while nursing;  (2d) Other specific use restrictions as noted.

*Stillingia sylvatica* Garden ex L.          Euphorbiaceae
**Common Name:** stillingia
**Other Common Names:** queen's-delight, queen's root, yaw root
**Part:** root
**Class:** 2c [30]
**Ed. Note:** Fresh root contains caustic latex and may be irritating to mucosa [10].

*Styrax benzoin* Dryander.          Styracaceae
*Styrax paralleloneurum* Perkins.
*Styrax tonkinensis* (Pierre) Craib. et Hartw.
**Common Name:** benzoin
**Other Common Names:** Sumatra benzoin
**Part:** gum resin
**Class:** 1

*Symphytum asperum* Lepechin.          Boraginaceae
*Symphytum x uplandicum* Nyman.
**Common Name:** Russian comfrey
**Part:** leaf, root
**Class:** 2a [8, 30]; 2d - long-term use is not recommended [30].
**Notice:** Pyrrolizidine Alkaloids (0.01-0.15% in leaves; up to 0.37% in root) [1, 3, 8, 17] *See page 149.*
**Ed. Note:** Effective July 1996, the AHPA Board of Trustees recommends that all products with botanical ingredient(s) containing toxic pyrrolizidine alkaloids, which include S. *asperum* and S. x *uplandicum*, display the following cautionary statement on the label:

"For external use only. Do not apply to broken or abraded skin. Do not use when nursing."

These species contain the pyrrolizidine alkaloid echimidine [5, 8], one of the most toxic of these compounds.

Australian regulations prohibit use of S. x *uplandicum* except when contained in homeopathic preparations in dilutions of 0.000001% or greater [1]. Canadian regulations do not allow its use in foods [3].

*Symphytum officinale* L.          Boraginaceae
**Common Name:** comfrey
**Other Common Names:** common comfrey, healing-herb
**Part:** leaf, root

---

Class 3: Herbs for which significant data exist to recommend the following labeling: "To be used only under the supervision of an expert qualified in the appropriate use of this substance." Labeling must include proper use information: dosage, contraindications, potential adverse effects and drug interactions, and any other relevant information related to the safe use of this substance.
Class 4: Herbs for which insufficient data are available for classification.

**Class:** 2a [4, 5, 8, 27, 30]; 2b [8]; 2c [8]

**Standard Dose:** External application should be limited to 4-6 weeks at a daily dosage at or below 100 µg unsaturated pyrrolizidine alkaloids [4, 27].

**Notice:** Pyrrolizidine Alkaloids (0.012-0.16% in leaf; 0.3-0.4% in dried root) [1, 3, 4, 5, 8, 17, 21, 24, 26, 27, (Hobbs, 1991a)] *See page 149.*

**Ed. Note:** Effective July 1996, the AHPA Board of Trustees recommends that all products with botanical ingredient(s) containing toxic pyrrolizidine alkaloids, which include *Symphytum officinale*, display the following cautionary statement on the label:

"For external use only. Do not apply to broken or abraded skin. Do not use when nursing."

Considering the lower content of alkaloids in the leaf of *S. officinale* and the absence of echimidine in most samples of this species, the limitation to external use recommended above may be overly cautious for the leaf. De Smet [8], however, without differentiation of species, states that internal use of *Symphytum* preparations may cause severe hepatic damage in chronic use. Internal use for more than 4-6 weeks per year is, in any event, discouraged.

Adulteration of *Symphytum officinale* may occur with *Symphytum* species which contain echimidine (one of the most toxic of the pyrrolizidine alkaloids), such as *S. asperum* and *S. x uplandicum*) [27].

Also, gas chromatographic analysis has reported echimidine to be present in one quarter of tested samples of *S. officinale*, in a range reported as from a "...generally very small quantity..." up to "...a rather high echimidine content comparable to concentrations found in *Symphytum asperum*..." (Awang et al., 1993).

Australian regulations prohibit use except when contained in homeopathic preparations in dilutions of 0.000001% or greater [1]. Canadian regulations do not allow comfrey in foods [3].

## *Symplocarpus foetidus* (L.) Salisb. ex Nutt.     Araceae

**Common Name:** skunk cabbage
**Part:** root, herb
**Class:** 2d - Individuals with a history of kidney stones should use this herb cautiously [30].
**Notice:** GI Irritant [10] *See page 172.*
Oxalates [15, 16, 18] *See page 148.*
**Ed. Note:** Fresh root is irritating to mucosa [30].

---

Class 1: Herbs that can be safely consumed when used appropriately.
Class 2: Herbs for which the following use restrictions apply, unless otherwise directed by an expert qualified in the use of the described substance:
    (2a) For external use only;      (2b) Not to be used during pregnancy;
    (2c) Not to be used while nursing;      (2d) Other specific use restrictions as noted.

*Syzygium aromaticum* (L.) Merr. & L. M. Perry.  Myrtaceae
Common Name: clove
Part: bud
Class: 1


*Syzygium cumini* Skeels.  Myrtaceae
Common Name: jambolan
Other Common Names: jumbul
Part: bark, seed
Class: 1


*Tabebuia heptaphylla* (Vell.) Toledo.  Bignoniaceae
*Tabebuia impetiginosa* (Mart. ex DC.) Standl.
Common Name: pau d'arco
Other Common Names: lapacho colorado (*T. heptaphylla*), lapacho morado (*T. heptaphylla*), ipe roxo (*T. heptaphylla, T. impetiginosa*), taheebo (*T. heptaphylla, T. impetiginosa*)
Part: bark
Class: 1


*Tanacetum parthenium* (L.) Schultz-Bip.  Asteraceae
Synonym: *Chrysanthemum parthenium* (L.) Berhn.
Common Name: feverfew
Part: herb
Class: 2b [8, (Mitchell et al., 1983)]
Ed. Note: Occasional side effects, such as mouth ulceration or gastric disturbance, have been observed in from 6-15% of users, usually in the first week of use [5, 8, 21, 28]. There are no known adverse effects in long-term consumption [5].


*Tanacetum vulgare* L.  Asteraceae
Synonym: *Chrysanthemum vulgare* (L.) Berhn.
Common Name: tansy
Part: herb
Class: 2b [6, 18, 30, (Osol & Farrar, 1955)]; 3 [1, 12, 18, 19]
Notice: Abortifacient [12, (Osol & Farrar, 1955)] *See page 163.*
Emmenagogue/Uterine Stimulant [6, 10] *See page 169.*
Thujone in some chemotypes (0.14-0.42%) [6, 18, (Mitchell et al., 1983)] *See page 158.*

---

Class 3: Herbs for which significant data exist to recommend the following labeling:
"To be used only under the supervision of an expert qualified in the appropriate use of this substance." Labeling must include proper use information: dosage, contraindications, potential adverse effects and drug interactions, and any other relevant information related to the safe use of this substance.
Class 4: Herbs for which insufficient data are available for classification.

**Ed. Note:** The thujone content in *Tanacetum* is variable, leading to concern even within normal dosage range [4, 18].

*Schedule 705* of the Canadian Food and Drug Regulations lists the oil as allowed in alcoholic beverages only if thujone-free [3]. Regulated in the U.S. as an allowable flavoring agent in alcoholic beverages only, and only if thujone-free [20].

### *Taraxacum officinale* G. H. Weber ex Wigg. — Asteraceae
**Common Name:** dandelion
**Part:** leaf
**Class:** 1

**Part:** root
**Class:** 2d - Contraindicated in blockage of the bile ducts, acute gallbladder inflammation, and intestinal blockage [4, 5].
**Ed. Note:** Although two references report that superacidic stomach problems may occur from the use of *T. officinale* [4, 5], the language used to express this concern raises questions. Bradley [5], in discussing the leaf, declares that, "As with all bitter-containing drugs, hyperacidity can occur in the stomach." Commission E [4] uses almost identical wording in commenting on the side effects of the root and herb. This leads the editors to believe that our references have extrapolated from concerns which are relevant to other bitter herbs, such as gentian, resulting in a warning that is contrary to the experience of the editors.

### *Terminalia bellerica* (Gaertn.) Roxb. — Combretaceae
**Common Name:** Belleric myrobalan
**Other Common Names:** behada
**Part:** fruit
**Class:** 1
**Notice:** Tannins (up to 17.0%) [18] *See page 155.*

### *Terminalia chebula* (Gaertn.) Retz. — Combretaceae
**Common Name:** tropical almond, he zi
**Other Common Names:** Indian almond
**Part:** fruit
**Class:** 2d - Contraindicated in acute cough, acute diarrhea, and early stage dysentery [2, 29].
**Notice:** Tannins (25.0-30.0%) [18] *See page 155.*

---

Class 1: Herbs that can be safely consumed when used appropriately.
Class 2: Herbs for which the following use restrictions apply, unless otherwise directed by an expert qualified in the use of the described substance:
   (2a) For external use only;   (2b) Not to be used during pregnancy;
   (2c) Not to be used while nursing;   (2d) Other specific use restrictions as noted.

## *Ternstroemia pringlei* (Rose.).        Theaceae
Common Name: Tilia estrella
Part: flower
Class: 4


## *Teucrium chamaedrys* L.        Lamiaceae
Common Name: germander
Part: herb
Class: 3 (Kouzi et al., 1994; Larrey et al., 1994)
Ed. Note: Regulated in the U.S. as an allowable flavoring agent in alcoholic beverages only [20]. Canadian regulations do not allow germander as a non-medicinal ingredient for oral use products (Michols, 1995).

*Teucrium chamaedrys* has been associated with hepatotoxicity in humans (Mostefa-Kara et al., 1992), but another species (*Teucrium stocksianum*) demonstrated hepatoprotective effects in mice (Rasheed et al., 1995).


## *Teucrium scorodonia* L.        Lamiaceae
Common Name: wood sage
Part: herb
Class: 4
Ed. Note: *Teucrium chamaedrys* has been associated with hepatotoxicity in humans (Mostefa-Kara et al., 1992), but another species (*Teucrium stocksianum*) demonstrated hepatoprotective effects in mice (Rasheed et al., 1995). It is not known whether other species of *Teucrium* have similar effects.


## *Thuja occidentalis* L.        Cupressaceae
Common Name: thuja
Other Common Names: northern white cedar, eastern white cedar, eastern arborvitae, swamp cedar
Part: stem, frond
Class: 2b [1, 6, 18, 28, 30]; 2d - Not for long-term use; do not exceed recommended dose [1, 30].
Standard Dose: 2.0-4.0 ml of a liquid extract of unspecified concentration, which "should be taken internally only occasionally" [28].
Notice: Abortifacient [10, (Osol & Farrar, 1955)] *See page 163*.
Emmenagogue/Uterine Stimulant [6, 18, 28] *See page 169*.
Thujone (up to 0.65%) [6, 10, 17, 28] *See page 158*.
Ed. Note: Approved for food use in U.S. if thujone-free [20]. Canadian regulations allow thuja as a flavoring agent for alcoholic beverages if thujone-free [3].

---

Class 3: Herbs for which significant data exist to recommend the following labeling: "To be used only under the supervision of an expert qualified in the appropriate use of this substance." Labeling must include proper use information: dosage, contraindications, potential adverse effects and drug interactions, and any other relevant information related to the safe use of this substance.

Class 4: Herbs for which insufficient data are available for classification.

## *Thymus* x *citriodorus* (Pers.) Schreb.     Lamiaceae
**Common Name:** lemon thyme
**Part:** leaf
**Class:** 1


## *Thymus vulgaris* L.     Lamiaceae
**Common Name:** thyme
**Other Common Names:** garden thyme, common thyme
**Part:** herb
**Class:** 1
**Ed. Note:** The classifications and concerns for this herb are based upon therapeutic use and dosage and may not be relevant to its consumption as a spice.

Several authors list thyme as an emmenagogue or note that the essential oil should be avoided in pregnancy [6, 10, 25, 27].


## *Tilia* x *europaea* L.     Tiliaceae
**Common Name:** linden
**Other Common Names:** common European linden, lime leaves, European linden
**Part:** leaf, flower
**Class:** 1
**Ed. Note:** Regulated in the U.S. as an allowable flavoring agent in alcoholic beverages only [20].


## *Tilia platyphyllos* Scop.     Tiliaceae
**Common Name:** large leafed linden, lime tree flower
**Other Common Names:** linden
**Part:** leaf, flower
**Class:** 1
**Ed. Note:** Regulated in the U.S. as an allowable flavoring agent in alcoholic beverages only [20].


## *Trichosanthes kirilowii* Maxim.     Cucurbitaceae
**Common Name:** trichosanthes, *tian hua fen* (root), *gua lou* (fruit), *gua lou ren* (seed)
**Other Common Names:** Chinese cucumber, Chinese snakegourd
**Part:** fruit
**Class:** 1

---

Class 1: Herbs that can be safely consumed when used appropriately.
Class 2: Herbs for which the following use restrictions apply, unless otherwise directed by an expert qualified in the use of the described substance:
    (2a) For external use only;     (2b) Not to be used during pregnancy;
    (2c) Not to be used while nursing;     (2d) Other specific use restrictions as noted.

**Ed. Note:** Chang and But [7] report rare cases of mild diarrhea and gastric discomfort [7].

Part: root
Class: 2b [2, 23, 30]
Notice: Abortifacient [13, 23] *See page 163.*

Part: seed
Class: 1

*Trifolium pratense* L.      Fabaceae
Common Name: red clover
Part: herb, flower
Class: 2b [30]

*Trigonella foenum-graecum* L.      Fabaceae
Common Name: fenugreek, *hu lu ba*
Part: seed
Class: 2b [17, 29]

*Trillium erectum* L.      Liliaceae
Common Name: beth root
Other Common Names: birth root, wakerobin
Part: root
Class: 2b [22]
Notice: Emmenagogue/Uterine Stimulant [22] *See page 169.*
GI Irritant [30] *See page 172.*

*Turnera diffusa* Willd. ex Schult. var. *diffusa*      Turneraceae
Common Name: damiana
Part: leaf
Class: 1

*Tussilago farfara* L.      Asteraceae
Common Name: coltsfoot
Part: flower (*kuan dong hua*)
Class: 2b [8]; 2c [8]; 2d - long-term use is not recommended [30].
Notice: Pyrrolizidine Alkaloids (0.004-0.03%, calculated as senkirkine) [1, 4, 8, 24, 27, 28] *See page 149.*

---

Class 3: Herbs for which significant data exist to recommend the following labeling: "To be used only under the supervision of an expert qualified in the appropriate use of this substance." Labeling must include proper use information: dosage, contraindications, potential adverse effects and drug interactions, and any other relevant information related to the safe use of this substance.
Class 4: Herbs for which insufficient data are available for classification.

**Ed. Note:** The AHPA Board of Trustees recommends that all products with botanical ingredient(s) which contain toxic pyrrolizidine alkaloids, including *Tussilago farfara*, display the following cautionary statement on the label:

"For external use only. Do not apply to broken or abraded skin. Do not use when nursing."

*Tussilago* flowers of Chinese origin are reported by De Smet to usually contain a much larger amount of the toxic pyrrolizidine alkaloids [8]. Some texts, however, list the flowers of *Petasites japonica* as interchangeable with those of coltsfoot [13], a fact which may contribute to such quantitative disparity.

Australian regulations prohibit use except when contained in homeopathic preparations in dilutions of 0.000001% or greater [1]. Canadian regulations do not allow coltsfoot in foods [3].

**Part:** leaf
**Class:** 2b [4, 8]; 2c [4, 8]; 2d - do not exceed recommended dose [4]; not for long-term use [4, 27].
**Standard Dose:** 1.5 to 6 grams of the fresh or dried leaf or its equivalent in finished preparations; daily consumption of pyrrolizidine alkaloids with a 1-2 unsaturated necine structure must not exceed 10 µg for teas, or 1 µg for extracts and pressed juice from fresh leaf; limit use to not longer than 4-6 weeks per year [4, 27].
**Notice:** Pyrrolizidine Alkaloids (0.005%, primarily senkirkine) [1, 4, 8, 27, 28] *See page 149.*
**Ed. Note:** Alone among the plants containing toxic pyrrolizidine alkaloids, the leaf of coltsfoot is still allowed for internal use by the German Commission E [4].

## *Ulmus rubra* Muhl.                                                                 Ulmaceae
**Common Name:** slippery elm
**Part:** bark
**Class:** 1

## *Uncaria gambir* (Hunter) Roxb.                                                   Rubiaceae
**Common Name:** gambir
**Part:** leaf and twig
**Class:** 1
**Notice:** Tannins (calculated as catechin, 30.0-35.0%) [5, 13, 17, 18] *See page 155.*

---

Class 1: Herbs that can be safely consumed when used appropriately.
Class 2: Herbs for which the following use restrictions apply, unless otherwise directed by an expert qualified in the use of the described substance:
    (2a) For external use only;    (2b) Not to be used during pregnancy;
    (2c) Not to be used while nursing;    (2d) Other specific use restrictions as noted.

## *Uncaria tomentosa* (Wildd.) DC.     Rubiaceae
Common Name: cat's claw
Part: root
Class: 4
Notice: Tannins [30] *See page 155.*
Ed. Note: Ethnobotanical data are scanty. Jones reports that traditional usage includes use as a contraceptive, a history which suggests avoidance in pregnancy (Jones, 1995). The same author lists a wide range of contraindications, e.g., patients undergoing skin grafts and organ transplants; hemophiliacs prescribed fresh blood plasma; simultaneous administration of certain vaccines, hormone therapies, thymus extracts, and insulin; and children under three years of age.

## *Urtica dioica* L.     Urticaceae
Common Name: stinging nettle
Other Common Names: nettles
Part: leaf
Class: 1

## *Usnea barbata* (L.) Wigg. and other *Usnea* species     Usneaceae
Common Name: usnea
Other Common Names: tree moss, usnea lichen, old man's beard, beard moss
Part: lichen
Class: 1

## *Vaccinium angustifolium* Aiton.     Ericaceae
## *Vaccinium corymbosum* L.
## *Vaccinium pallidum* Aiton.
Common Name: blueberry
Part: leaf
Class: 1

## *Vaccinium myrtillus* L.     Ericaceae
Common Name: bilberry
Other Common Names: whortleberry, huckleberry
Part: leaf
Class: 4

---

Class 3: Herbs for which significant data exist to recommend the following labeling: "To be used only under the supervision of an expert qualified in the appropriate use of this substance." Labeling must include proper use information: dosage, contraindications, potential adverse effects and drug interactions, and any other relevant information related to the safe use of this substance.
Class 4: Herbs for which insufficient data are available for classification.

**Ed. Note:** The potential for significant side effects associated with chronic high doses has been discussed by some authorities [4, 27]. A review of the existing literature by De Smet [9] concluded that "...it is impossible to come up with a final verdict on the toxic potential of *Vaccinium myrtillus* leaves without the aid of new toxicological investigations."

Part: fruit
Class: 1

## *Valeriana edulis* Nutt. ex Torr. & Gray ssp. *procera* (Kunth) F.G. Mey.     Valerianaceae
## *Valeriana officinalis* L.
## *Valeriana sitchensis* Bong.
## *Valeriana wallichii* DC.

Common Name: valerian
Other Common Names: Mexican valerian (*V. edulis*), Pacific valerian (*V. sitchensis*), Indian valerian (*V. wallichii*), garden heliotrope (*V. officinalis*), garden valerian (*V. officinalis*)
Part: root, rhizome
Class: 1

**Ed. Note:** Although reports of toxicity of the valepotriates have been published, poor absorption and quick degradation into less toxic metabolites are so slight as to assure no acute adverse reactions [5, 17].

Required labeling for a specific registered valerian tincture in Germany warns of reduced ability to drive or operate machinery [5], although Van Hellemont [24] specifically states that no impairments in concentration are associated with valerian consumption.

"...There is some concern about continual use, which may cause minor side effects, including headaches, excitability, and insomnia." (Morazzoni & Bombardelli, 1995).

## *Vanilla planifolia* Andr.     Orchidaceae
## *Vanilla tahitensis* J. W. Moore.

Common Name: vanilla
Other Common Names: Bourbon vanilla (*V. planifolia*), Mexican vanilla (*V. planifolia*), Madagascar vanilla (*V. planifolia*), Tahitian vanilla (*V. tahitensis*)
Part: fruit
Class: 1

---

Class 1: Herbs that can be safely consumed when used appropriately.
Class 2: Herbs for which the following use restrictions apply, unless otherwise directed by an expert qualified in the use of the described substance:
    (2a) For external use only;    (2b) Not to be used during pregnancy;
    (2c) Not to be used while nursing;    (2d) Other specific use restrictions as noted.

***Veratrum viride* Aiton.**                          **Liliaceae**
Common Name: American hellebore
Other Common Names: American white hellebore, Indian poke, false hellebore
Part: root
Class: 3 [1, 10, 12, 15, 16, 18, 19, 21, 22, 26, 28]

***Verbascum thapsus* L.**                           **Scrophulariaceae**
***Verbascum densiflorum* Bertoloni**
***Verbascum phlomoides* L.**
Common Name: mullein
Part: leaf, flower
Class: 1
Ed. Note: Flowers are regulated in the U.S. as an allowable flavoring agent in alcoholic beverages only [20].

***Verbena hastata* L.**                                  **Verbenaceae**
Common Name: blue vervain
Part: herb
Class: 2b [27]

***Verbena officinalis* L.**                               **Verbenaceae**
Common Name: European vervain
Part: herb
Class: 2b [14, 29]
Ed. Note: Regulated in the U.S. as an allowable flavoring agent in alcoholic beverages only [20].

***Veronica officinalis* L.**                           **Scrophulariaceae**
Common Name: speedwell
Part: herb
Class: 1
Ed. Note: Regulated in the U.S. as an allowable flavoring agent in alcoholic beverages only [20].

***Vetiveria zizanoides* (L.) Nash.**                     **Poaceae**
Common Name: vetiver
Part: root

---

Class 3: Herbs for which significant data exist to recommend the following labeling: "To be used only under the supervision of an expert qualified in the appropriate use of this substance." Labeling must include proper use information: dosage, contraindications, potential adverse effects and drug interactions, and any other relevant information related to the safe use of this substance.
Class 4: Herbs for which insufficient data are available for classification.

**Class:** 2b [18, 25, 30]
**Notice:** Abortifacient [25] *See page 163.*
Emmenagogue/Uterine Stimulant [18] *See page 169.*
**Ed. Note:** Regulated in the U.S. as an allowable flavoring agent in alcoholic beverages only [20].

### *Viburnum opulus* L.      Caprifoliaceae
**Common Name:** cramp bark
**Other Common Names:** guelder rose
**Part:** bark, root bark
**Class:** 1
**Ed. Note:** Canadian regulations do not allow cramp bark as a non-medicinal ingredient for oral use products (Michols, 1995).

### *Viburnum prunifolium* L.      Caprifoliaceae
**Common Name:** black haw
**Part:** bark
**Class:** 2d - Individuals with a history of kidney stones should use this herb cautiously [30].
**Notice:** Oxalates [17] *See page 148.*

### *Vinca minor* L.      Apocynaceae
**Common Name:** common periwinkle
**Other Common Names:** periwinkle
**Part:** herb
**Class:** 2d - Contraindicated with low blood pressure [18] and in constipation (Mitchell et al., 1983).

### *Viola odorata* L.      Violaceae
**Common Name:** sweet violet
**Other Common Names:** violet
**Part:** leaf
**Class:** 1

### *Viola sororia* Willd.      Violaceae
**Common Name:** violet
**Other Common Names:** woolly blue violet
**Part:** leaf
**Class:** 4

---

Class 1: Herbs that can be safely consumed when used appropriately.
Class 2: Herbs for which the following use restrictions apply, unless otherwise directed by an expert qualified in the use of the described substance:
    (2a) For external use only;      (2b) Not to be used during pregnancy;
    (2c) Not to be used while nursing;      (2d) Other specific use restrictions as noted.

*Viola tricolor* L.  Violaceae
**Common Name:** heartsease
**Other Common Names:** European wild pansy, Johnny jump up
**Part:** herb
**Class:** 1
**Ed. Note:** Regulated in the U.S. as an allowable flavoring agent in alcoholic beverages only [20].

*Viscum album* L.  Viscaceae
**Common Name:** European mistletoe
**Part:** herb
**Class:** 2d - Contraindicated in protein hypersensitivity and chronic-progressive infections such as tuberculosis and AIDS [4, 11, 24]; do not exceed recommended dose [1, 30].
**Standard Dose:** 2.5 grams, infused in cold water for 10 to 12 hours, up to two times daily [27].
**Notice:** Lectins [17, 21, 27] *See page 146.*
**Ed. Note:** It is advised that blood pressure should be checked regularly by those consuming the tea [27].

Although Leung & Foster [17] report that the majority of recorded ingestions show no toxic symptoms, this reference advises against self-medication. Side effects related to injections are recorded, including chills, fever, headache, symptoms of angina, circulatory problems, and allergic reactions [4].

Canadian regulations do not allow European mistletoe in foods [3]. The following labeling has been proposed for products sold in Australia: "Do not take with medications prescribed for high or low blood pressure except on professional advice" [1].

*Vitex agnus-castus* L.  Verbenaceae
**Common Name:** chaste tree
**Part:** fruit (berry)
**Class:** 2b [4, (Hobbs, 1990)]; 2d - May counteract the effectiveness of birth control pills (Hobbs, 1990).
**Notice:** Emmenagogue/Uterine Stimulant [30] *See page 169.*
**Ed. Note:** Occasional minor skin irritations have been reported [4, 17].

As an amphoteric hormone-regulating herb, *Vitex* is recommended for slowing the menstrual flow and reducing spotting [18], as well as an emmenagogue to induce menstrual flow [Madaus, 1938], depending on the

---

Class 3: Herbs for which significant data exist to recommend the following labeling: "To be used only under the supervision of an expert qualified in the appropriate use of this substance." Labeling must include proper use information: dosage, contraindications, potential adverse effects and drug interactions, and any other relevant information related to the safe use of this substance.
Class 4: Herbs for which insufficient data are available for classification.

circumstances and hormone status of specific users.

Although generally not recommended for use in pregnancy, *Vitex* has been used to prevent miscarriage in the first trimester of pregnancy with cases of progesterone insufficiency [30].

### *Withania somnifera* Dunal. — Solanaceae
**Common Name:** ashwagandha
**Part:** root
**Class:** 2b [6, 25, 30]; 2d - May potentiate the effects of barbiturates (Atal and Schwarting, 1961).
**Notice:** Abortifacient [6, 25] *See page 163.*

### *Wolfiporia cocos* (Schwein.) Ryv. & Gilbn. — Polyporaceae
**Common Name:** polyporus, *fu ling, fu shen*
**Part:** sclerotium
**Class:** 1

### *Yucca aloifolia* L. — Agavaceae
### *Yucca brevifolia* Engelm.
### *Yucca glauca* Nutt.
### *Yucca whipplei* Torr.
**Common Name:** yucca
**Other Common Names:** aloe yucca (*Y. aloifolia*), Spanish bayonet (*Y. aloifolia*), dagger plant (*Y. aloifolia*), Joshua tree (*Y. brevifolia*), our Lord's candle (*Y. whipplei*), soapweed (*Y. glauca*)
**Part:** root
**Class:** 1

### *Zanthoxylum americanum* Mill. — Rutaceae
### *Zanthoxylum clava-herculis* L.
**Common Name:** prickly ash
**Other Common Names:** toothache tree, northern prickly ash, southern prickly ash, Hercules' club
**Part:** bark
**Class:** 2b [5]

---

Class 1: Herbs that can be safely consumed when used appropriately.
Class 2: Herbs for which the following use restrictions apply, unless otherwise directed by an expert qualified in the use of the described substance:
    (2a) For external use only;    (2b) Not to be used during pregnancy;
    (2c) Not to be used while nursing;    (2d) Other specific use restrictions as noted.

*Zanthoxylum simulans* Hance.　　　　　　　　　　Rutaceae
*Zanthoxylum bungeanum* Maxim.
*Zanthoxylum schinifolium* Sieb.
Common Name: Sichuan pepper, *chuan jiao*
Other Common Names: Szechuan pepper
Part: fruit rind
Class: 2b [2, 29]


*Zea mays* L.　　　　　　　　　　　　　　　　　Poaceae
Common Name: corn, cornsilk
Part: stigma
Class: 1
Ed. Note: Regulated in the U.S. as an allowable ingredient in certain listed foods in a range of from 4.0 to 30.0 ppm [20].


*Zingiber officinale* Roscoe.　　　　　　　　　　Zingiberaceae
Common Name: ginger
Part: fresh root
Class: 1

Part: dried root
Class: 2b [4, 29]; 2d - Persons with gallstones should consult a practitioner prior to use [4].
Ed. Note: The classifications and concerns for this herb are based upon therapeutic use and may not be relevant to its consumption as a spice.


*Ziziphus jujuba* Mill.　　　　　　　　　　　　Rhamnaceae
Common Name: jujube, *da zao*
Other Common Names: Chinese jujube
Part: fruit
Class: 1


*Ziziphus spinosa* Hu.　　　　　　　　　　　　Rhamnaceae
Common Name: jujube seeds, *suan zao ren*
Part: seed
Class: 2b [17, 30]
Notice: Emmenagogue/Uterine Stimulant [17] *See page 169.*

---

Class 3: Herbs for which significant data exist to recommend the following labeling: "To be used only under the supervision of an expert qualified in the appropriate use of this substance." Labeling must include proper use information: dosage, contraindications, potential adverse effects and drug interactions, and any other relevant information related to the safe use of this substance.
Class 4: Herbs for which insufficient data are available for classification.

# Introduction to Appendices

The first two appendices which follow are included to provide a more thorough understanding of some specific safety issues which are relevant to the consumption of certain plants; the third appendix is included as a ready to use reference to the plants included in cautionary classes.

In order to present meaningful safety classifications of many of the plants included in this work, a **Notice** is included at the plant's listing. There are fifteen **Notices** for constituents, such as caffeine or pyrrolizidine alkaloids, and twelve additional **Notices** for actions associated with particular plants, such as emetics and nervous system stimulants.

*Appendix 1* includes a basic overview of each of the constituent **Notices**, presented with a discussion of the adverse effects and therapeutic uses of both the constituent and the plants which contain it. Pharmacological data are also included, as well as a list of each of the plants covered in the text and referenced for the particular constituent.

*Appendix 2* addresses the **Notices** for physiological actions which are associated with some herbs and which are a cause for some concern. Adverse effects and therapeutic use are also described for these categories, and the mechanism of action is presented in most cases. Again, a list of the plants which are referenced for the identified concern is included.

It is important to understand the cursory nature of the abstracts presented in these appendices. Their inclusion here is intended only to provide sufficient understanding of the issues involved with each of these constituents and actions to assure that consumers will be informed regarding potential adverse effects. For a more thorough understanding of these, the references cited in each section should be consulted.

*Appendix 3* provides listings of all of the herbs included in each of four of the **Classes** established in the *Botanical Safety Handbook*. Separate lists are provided of common names and botanical names for four of the five restrictive **Classes**. Class 2d was excluded solely because such a list would not be meaningful due to the general nature of its definition. Each of these lists should be seen as defining only those plants which are included in the *Handbook*, and not as an exhaustive presentation.

# Appendix 1

## HERBAL CONSTITUENT PROFILES

## Aristolochic Acid

### Abstract

Aristolochic acid is principally found in members of the Aristolochiaceae family. It consists of a group of related nitrophenanthrene carboxylic acid derivatives and appears in two main forms: aristolochic acid I (AAI) and aristolochic acid II (AAII). Both of these have extremely potent carcinogenic, mutagenic, genotoxic, and nephrotoxic potential in a number of animal models. Nephrotoxicity has also been observed in humans. The chemical structure is 3,4-methylenedioxy-8-methoxy-10-nitrophenanthrene-1-carboxylic acid.

### Adverse Effects

Because aristolochic acid is both carcinogenic and nephrotoxic, the plants containing it are not recommended for long-term use. Even though there is an extensive history of use of these plants with no reports of carcinogenic effects, recent events regarding weight loss preparations which contain herbs high in aristolochic acid have raised concerns over toxicity in humans.

Cases of nephrotoxicity were reported from ingestion of weight loss products that listed *Stephania tetranda* and *Magnolia officinalis* as ingredients. When analyzed, these products were found to contain aristolochic acid (Vanhaelen et al., 1994), a constituent which does not occur in either of these plants. It is now clear that the product actually contained the herb *Aristolochia fangchi*, which is commonly substituted for *Stephania* in Chinese herbal medicine (Bensky & Gamble, 1986; De Smet, 1993; Vanderweghem, 1994). The reported nephrotoxic effects consisted of a unique type of rapidly progressive renal fibrosis. Analysis of the biopsied necrosis confirmed the presence of aristolochic acid adducts to the nephronal DNA.

Long-term oral administration of pure AAI to male rats induced multiple tumors, mainly in the forestomach, glandular stomach, ear duct, and small intestine. Adenomas of the kidneys, carcinomas of the lungs, and haemangiomas of the uteri were also seen in mice after 3 weeks of treatment with aristolochic acid at a dosage level of 5.0 mg/kg (Mengs & Klein, 1988). With acute toxicity, the $LD_{50}$ ranged from 56 to 203 mg/kg or 38 to 83 mg/kg intravenously (Mengs, 1987).

A human study demonstrated *in vitro* aristolochic acid induction of structural chromosome aberrations and sister chromatid exchanges in human lymphocytes. The results showed a direct correlation between dose and the induction of gaps and breaks (Abel & Schimmer, 1983).

### Therapeutic Use

Plants containing aristolochic acid have long been used as anti-inflammatory agents and herbal drugs in many cultures. Various species of plants from the family Aristolochiaceae, e.g., Virginia snakeroot (*Aristolochia serpentaria*), as well as Canada snakeroot (*Asarum canadense*), are mentioned in the ethnobotanical literature of North America for the treatment of snakebites.

In 1982, aristolochic acid was shown to be a mutagen and strong carcinogen in rats. Plants containing aristolochic acid are no longer recommended for human consumption in Germany and other countries (Abel & Schimmer, 1983; Pezzuto et al., 1986).

### Pharmacology

In animal tests, aristolochic acid has proven to be anti-inflammatory, antiviral, a stimulant to immune system activity, carcinogenic, genotoxic, and mutagenic. It has demonstrated viral protective effects in a double blind test with rabbits (Mose et al., 1980). An immune stimulating action has been reported in a human study that measured increased phagocytic activity after three days of treatment at a dosage of 0.9 mg/day (Kluthe et al., 1982).

Aristolochic acid has been shown to bind to surface receptors of lymphocytes, altering immune response (Siering & Muller, 1981). It remains in the tissues for an extended length of time, mostly in a bound form as DNA adducts. A study with rats found detectable amounts of aristolochic acid nine months following an initial single dose. The primary target organ of aristolochic acid is the forestomach. The glandular stomach, liver, lung, and urinary bladder are secondary targets (Fernando et al., 1993).

Aristolochic acid also plays a regulatory role in prostaglandin synthesis. It has an anti-inflammatory effect, inhibiting inflammation by immunological and nonimmunological agents. One mechanism of activity may be as a direct blocking/protein-targeted inhibitor of phospholipase A2, decreasing the generation of eicosanoids and platelet-activating factors that are responsible for edema. Another anti-inflammatory mechanism may be aristolochic acid's effect on arachidonic acid mobilization in human neutrophils (Moreno, 1993; Rosenthal et al., 1992).

**Herbs** listed in the *Botanical Safety Handbook* that contain aristolochic acid:

*Aristolochia clematitis*  
*Aristolochia contorta*  
*Aristolochia debilis*  
*Aristolochia serpentaria*  
*Asarum canadense*

# Atropine

## Abstract

Atropine and its racemate, hyoscyamine, are alkaloids with great potential toxicity that are found in several plants in the nightshade family, Solanaceae. Plants containing atropine are used principally for their antispasmodic properties.

## Adverse Effects

The atropine-containing herbs are not commonly found in trade. The primary toxicity concern for these is related to their occasional occurrence as adulterants. Even therapeutic doses of these plants are considered potentially toxic, and overdoses can be fatal. Atropine is known to block the action of the parasympathetic branch of the autonomic nervous system, producing in overdoses such observable symptoms as mental confusion, memory loss, constipation, difficult urination, dryness of the mouth and mucous membranes, nervousness, restlessness, and hallucinations.

Required labeling in the United States for atropine-containing drugs, including preparations from plant material, warns against unsupervised use by persons with glaucoma or excessive eye pressure, by children under 6, and by the elderly. Discontinuation of use is suggested if symptoms of blurred vision, rapid pulse, or dizziness are observed.

## Therapeutic Use

The four major medicinal plants that contain atropine and hyoscyamine are belladonna (*Atropa belladonna*), henbane (*Hyoscyamus niger*), mandrake (*Mandragora officinarum*), and thornapple, or jimson weed (*Datura stramonium*). Belladonna, henbane, and mandrake are native to Europe. *Datura* is indigenous to the Americas and is used by the native peoples of North America and Mexico for hallucinogenic and ritual purposes. The dried leaves of several of these plants are smoked for relief of bronchial asthma. Each of these species is cultivated by pharmaceutical companies as a source of alkaloids and as a starting material for synthetic drugs.

Atropine and hyoscyamine are therapeutically important and form the starting material for a number of commonly prescribed pharmaceutical drugs. The alkaloids and plants that contain them are considered antispasmodics. They reduce spasms in smooth muscles of the body, including the bladder, bronchial tree, and digestive tract. These plants and their purified alkaloids are prescribed as drugs to relieve cramps or spasms of the stomach, intestines, and bladder. They may also be used to treat peptic ulcers and to prevent nausea, vomiting, and motion sickness.

Atropine sulfate is official in the *United States Pharmacopeia*; belladonna herb is official in a number of pharmacopeias worldwide.

## Pharmacology

Atropine and hyoscyamine are tropane alkaloids. There are approximately 200 different tropane alkaloids known in the plant kingdom, primarily in the nightshade family (Solanaceae). Belladonna contains from 0.3-0.6%, henbane 0.04-0.15%, and thornapple 0.2-0.5% total alkaloids, mostly hyoscyamine. Belladonna and thornapple have nearly identical actions, therapeutic properties, and toxic potential; henbane is weaker but contains more of the sedative compound scopolamine.

Atropine acts on muscarinic receptors to block parasympathetic effects on smooth muscle, cardiac muscle, and glandular cells. It blocks vagal nerve activity, increasing firing rate of the sinoarial node. Pharmacological effects of atropine in clinical doses include reduced heart rate and peristalsis, increased bladder pressure, and relaxation of the gallbladder duct. Production of saliva, sweat, and gastric fluids as well as pancreatic, bronchial, and eye secretions are all reduced.

**Herbs** listed in the *Botanical Safety Handbook* that contain atropine:

*Atropa belladonna*                    *Mandragora officinarum*

## $\beta$-Asarone

### Abstract

$\beta$-Asarone is a potential hepatocarcinogenic constituent found in the essential oils of several plants in the *Acorus* and *Asarum* genera of the Araceae and Aristolochiaceae families, respectively. The compound is in a chemical group known as phenylpropanoids (C6-C3); more specifically, it is an allylbenzene, also known as an alkenylbenzene, or allylphenol. The main structural feature of the asarones ($\alpha$ and $\beta$) is an aromatic ring with a 2-methoxy group.

### Adverse Effects

The potential hazard to humans of low doses of allylbenzenes (e.g., $\beta$-asarone, estragole, and safrole) is very minimal. Consuming several grams will generate very small quantities of genotoxic metabolites that are quickly broken down by the cytosolic and microsomal epoxide hydrolases of the liver. Nevertheless, herbs containing $\beta$-asarone should not be used long-term because they have been documented to have chromosome damaging effects on human lymphocytes, mutagenic property in bacteria, and carcinogenic activity in rats (De Smet, 1992).

Studies demonstrating carcinogenic activity in animals were conducted with rodents fed or injected very high doses of $\beta$-asarone. In one of these studies, rats developed mesenchymal tumors of the small intestine (Keeler & Tu, 1983). Similar research demonstrated an increase of unscheduled DNA synthesis, a strong indicator of impending genotoxicity (Tsai et al., 1994). Another study showed that β-asarone had an anticoagulant effect in mice and rats (Rubio-Poo et al., 1991).

Certain varieties of *Acorus calamus* have the highest potential for adverse effects due to β-asarone exposure, while other varieties of the same plant are asarone-free. The following table summarizes the content of β-asarone in the essential oil of the three most commonly used varieties (Bruneton, 1995; Wichtl, 1994).

| Variety: | Range: | % β-asarone: | Chromosome Sets: |
|---|---|---|---|
| *americanus* | North America | absent | 2n |
| *calamus* | Europe | <10% | 3n |
| *angustatus* | India | up to 80-96% | 4n |

Although the American (diploid) and European (triploid) varieties of *Acorus calamus* are reported to contain at most only traces of β-asarone, careful attention should be paid to the identity of these because of the possibility of adulteration with the Indian variety or other variants containing unacceptable levels of the compound.

Short-term use of β-asarone-containing herbs in sufficient quantity may cause nausea and vomiting. All varieties of calamus are prohibited in foods in the United States and are listed as unacceptable non-medicinal ingredients for oral use in Canada (Office of Federal Register, 1994; Michols, 1995).

## Therapeutic Use

The use of calamus in Asia, Europe, and North America is longstanding. *Acorus* species have been official in many pharmacopeias and are now mainly used as sources of calamus oil, which is employed in perfumery (Trease & Evans, 1978) and for its antispasmodic, carminative, and digestive stimulating effects (Bruneton, 1995; Reynolds, 1989). The antispasmodic activity of the different varieties is proportional to their relative β-asarone levels, with spasmolytic activity decreasing with higher β-asarone concentrations (De Smet, 1992).

## Pharmacology

β-asarone is a procarcinogen that is neither hepatotoxic nor directly hepatocarcinogenic. It must first undergo metabolic 1'-hydroxylation in the liver before achieving toxicity. Cytochrome P450 in the hepatocytes is responsible for secreting the hydrolyzing enzymes that convert β-asarone into its genotoxic epoxide structure. Even with activation of these metabolites, the carcinogenic potency is low because of the rapid breakdown of epoxide residues with hydrolase that leaves the compounds inert (Luo, 1992). In addition, the major metabolite of β-asarone is 2,4,5-trimethoxycinnamic acid, a derivative that is not carcinogenic (Hasheminejad & Caldwell, 1994).

The activation of the procarcinogen β-asarone is different from that of the allylbenzene estragole and the propenylbenzene safrole. Asarone has a novel activation featuring hydroxylation of the 2-methoxy group of the aromatic ring.

**Herbs** listed in the *Botanical Safety Handbook* that contain β-asarone:

| | |
|---|---|
| *Acorus calamus* | *Asarum canadense* |
| *Acorus gramineus* | *Asarum europaeum* |

## *Berberine*

### Abstract

Berberine is a quaternary alkaloid ($C_{20}H_{19}NO_5$) found in the Berberidaceae, Papaveraceae, Ranunculaceae, Rutaceae, and other families. The alkaloid is bright yellow in appearance and is more specifically a quaternary ammonium protoberberine. Berberine has bitter characteristics and is used in traditional medicines for its anti-inflammatory and antimicrobial effects, among other things. The compound is moderately toxic and can depress the function of the heart.

### Adverse Effects

Berberine-containing herbs are not recommended for use during pregnancy. Although there is no specific study that has documented ill effects in pregnancy, several of these herbs have a recorded history of use as antifertility agents. Paradoxically, berberine, and in some cases its salts, have exhibited uterine contracting actions as well as anticonvulsive activity in laboratory animals (De Smet, 1992; Leung & Foster, 1996).

Pure berberine is considered moderately toxic ($LD_{50}$ in humans: 27.5 mg/kg), causing cardiac damage, dyspnoea, and lowered blood pressure in overdoses. Studies in India have attributed chronic wide-angle glaucoma associated with dropsy to sanguinarine, the berberine-like compound of bloodroot (Bruneton, 1995).

Long-term dog and cat studies (dosage: 2mg/kg of berberine) found berberine to depress the cardiac function through dilation of the blood vessels and stimulation of the vagus nerve. Sustaining this high dosage also depressed respiration, in addition to stimulating the smooth muscle of the intestines and uterus. Smaller doses were found to have an opposite effect, stimulating the cardiac and respiratory systems, while depressing peristalsis of the intestinal smooth muscle. Low amounts had a stimulatory effect on the heart muscle and increased the flow through the coronary artery (Osol & Farrar, 1955).

## Therapeutic Use

Berberine-containing herbs, especially from the genera *Mahonia*, *Berberis*, *Coptis*, and *Hydrastis* are extensively used in India, China, and other parts of Asia as well as North America. Therapeutic uses and actions of berberine-containing plants include antimalarial and antimicrobial functions. They are also used as antipyretics to control fevers, as anthelmintics to expel certain parasites, and as bitter stomachics to improve appetite and digestion (Harborne & Baxter, 1993).

*Coptis chinensis* and *Phellodendron amurense* are used in traditional Chinese medicine for "clearing heat" by relieving infection and lessening inflammation. Bloodroot (*Sanguinaria canadensis*), another berberine-containing medicinal plant, has made a significant contribution to the oral hygiene industry. A berberine-like constituent from bloodroot, in the form of sanguinarine chloride, is an effective inhibitor of oral bacteria and is used at low concentrations in toothpastes and mouthwashes (Bruneton, 1995). Bloodroot also has antimicrobial action against dysentery and gynecological infection.

The root and rhizome of the North American native goldenseal (*Hydrastis canadensis*) are used in bitter digestive tonics and immune system activating formulas, especially blended with members of the genus *Echinacea*. It was described in the Lewis and Clark journals of the early 19th century as a remedy employed by certain native American tribes as a wash to relieve eye irritation. Goldenseal and its alkaloid hydrastine continue to have utilization as ingredients in eyedrops for the treatment of conjunctivitis (Bruneton, 1995; Lewis & Elvin-Lewis, 1977). This versatile plant is also employed as a uterine tonic, acting to compress the uterine blood vessels and contract the uterine wall.

## Pharmacology

Berberine produces a variety of unique pharmacological effects, exhibiting antimicrobial, diuretic, smooth muscle relaxant, and cardiac depressant activities. It is reported to be bacteriostatic at low doses and bactericidal at higher doses and has been shown *in vitro* to be active against many organisms. Plants containing berberine have historically been used to fight a number of infectious organisms, and the sulfate, hydrochloride, and chloride forms are used in western pharmaceutical medicine as antibacterial agents (Bruneton, 1995; Reynolds, 1993).

Berberine has bitter characteristics and stimulates bile secretion, providing the rationale for the inclusion of these herbs in digestive tonics. Compounds closely related to berberine as well as berberine itself have cholekinetic action, as demonstrated by sanguinarine, a berberine relative which acts as an acetylcholinesterase inhibitor (Wichtl, 1994).

*Herbs* listed in the *Botanical Safety Handbook* that contain berberine:

| | |
|---|---|
| *Berberis vulgaris* | *Mahonia nervosa* |
| *Chelidonium majus* | *Mahonia repens* |
| *Coptis chinensis* | *Phellodendron amurense* |
| *Coptis groenlandica* | *Phellodendron chinense* |
| *Hydrastis canadensis* | *Sanguinaria canadensis* |
| *Mahonia aquifolium* | |

# *Cardiac Glycosides*

## Abstract

Cardiac glycosides are a class of steroid glycosides that specifically affect the function and rhythm of the heart muscle. They are primarily found in the Asclepiadaceae and Apocynaceae families and are scattered throughout a few genera in the Brassicaceae, Fabaceae, Liliaceae, Ranunculaceae, Scrophulariaceae, and Tiliaceae families. The *Digitalis* genus with its twenty species is the chief source of cardiac glycosides for the manufacture of pharmaceutical grade drugs, with some species of *Strophanthus* providing the starting material for others (Bruneton, 1995; Gilman et al., 1985).

## Adverse Effects

Intoxication from cardiac glycosides is common and hazardous. Of the four genera profiled in this work as containing these constituents, only *Asclepias* is acceptable for unsupervised consumption. None of the plants listed here, and certainly none of the chief drug source plants, are readily available as ingredients in consumer products.

In the hospital and clinical setting, adverse drug reactions to cardiac glycosides exceed all other prescribed drugs (Keller & Tu, 1983). As many as 25% of all patients exhibit some sign of toxicity (Gilman et al., 1985). They are particularly dangerous because the therapeutic dose is close to the toxic dose. Therapeutic blood plasma levels are considered to be from 0.5-2.0 ng per mL, while concentrations in excess of 1.5-2.0 ng per mL are described as an indication of risk (Reynolds, 1989). Consequently, all therapy must be individualized, and initial administration must be closely monitored. Also, the safety of long-term treatment with cardiac glycosides for chronic congestive heart failure is controversial, and a contemporary review of this issue has been initiated.

Outside the clinical environment, acute intoxication from cardiac glycoside containing herbs is seldom seen. Occasionally there are reports of lethal intoxication by individuals eating foxglove (*Digitalis* spp.) as a wild potherb, having mistaken its young leaves for those of comfrey. Although consumption of *Convallaria* can produce unpleasant gastrointestinal effects, serious intoxication is mitigated by the poor absorption of the contained glycoside and is reportedly rare (Bruneton, 1995). On the other hand, according to the *South African Medical Journal*, 44% of deaths caused by herbal medicines were due to those containing cardiac glycosides (Mcvann et al., 1992). The severity of accidental intoxication is predominantly related to the individual's underlying heart condition, as healthy persons are less likely to have severe arrhythmias than those with a preexisting heart condition.

Symptoms of mild to moderate acute intoxication from cardiac glycoside drugs as well as cardiac glycoside-containing plants include fatigue, anorexia, ventricular ectopic beats, bradycardia, nausea, headache, malaise, vomiting, seizures, hyperkalemia, and ventricular premature beats. Severe acute intoxication results in disorientation, confusion, nightmares, hallucinations, visual disturbances, diarrhea, ventricular tachycardia, and sinoatrial block that may lead to ventricular fibrillation. The most serious side effects of cardiac glycosides are those on the heart, such as supraventricular or ventricular arrhythmias, defects of cardiac conduction, and sinoatrial and atrioventricular block. These extreme effects on the heart are caused by the spontaneous release of calcium from the myocardium's sarcoplasmic reticulum. Symptoms of this degree of severity often terminate in death (Reynolds, 1993).

In the event of acute poisoning, the stomach must be emptied and activated charcoal given to reduce absorption of the cardiac glycoside. If absorption has occurred, treatment with arrhythmic drugs, atropine to prevent bradycardia, and/or the administration of F(ab) fragments of anti-digoxin (digitoxin; -oubain) antibodies should be considered. Antibody drugs provide the clinician with a rapidly acting, safe antidote for all commonly used digitalis preparations. Other drugs to prevent acute heart failure are membrane protective agents (unithiol, $\alpha$-tocopherol, hydrocortisone), pharmacological antagonists (novodrin, alupent, isoprenaline), and the cardiotonic agent dobutamine. Potassium chloride should also be administered, maintaining serum potassium levels greater than or equal to 4 mEq/liter to minimize the effects of cardiac glycosides on the electrolyte balance of the heart. An emergency medical staff should closely monitor intoxicated patients until the condition is stabilized (Reynolds, 1993; Kelly & Smith, 1992).

## Therapeutic Use

Medicinal plants containing cardiac glycosides, such as foxglove (*Digitalis* spp.) and squill (*Urginea maritima*), have been used for many centuries to treat heart conditions. Early records tell how these plants were used for headache, spasms, constipation, wound healing, and to induce vomiting. The genus *Digitalis*, for instance, was traditionally used by farmers as a diuretic to treat edema (Erdmann, 1986). Native cultures in Africa and Asia have used certain of the cardiac glycoside-containing plants in the manufacture of arrow poisons (Bruneton, 1995).

The earliest scientific and clinical publication on cardiac glycosides was written by William Withering in 1785. Withering was an English physician who learned of foxglove from a Shropshire herbalist who used the herb in the treatment of dropsy and subsequently observed its effects on several hundred patients over a ten-year period. His work not only gave digitalis a place in medicine as an important therapeutic agent, but served as a standard for testing and evaluating the performance of a drug (Erdmann, 1986). Since this time, cardiac glycosides from *Digitalis* and *Strophanthus* have remained the most important compounds in the treatment of congestive heart failure with pulmonary or systemic congestion, atrial flutter, and fibrillation.

The first purified cardiac glycoside, digitalin, was isolated from foxglove in 1868. Today, digoxin and digitoxin are extracted from the leaves of *Digitalis lanata* and *D. purpurea*, while ouabain is obtained from *Strophanthus*, and squill bulbs provide the source material for proscillaridin A, a cardiac glycoside used in Europe (Bruneton, 1995).

Digoxin is the most commonly prescribed of these drugs, accounting for more than 21 million prescriptions in the United States annually (Werbach & Murray, 1994). Because of its intermediate duration of action (36-hour half-life) and limited side effects, digoxin is generally preferred over other substances. Digitoxin, on the other hand, is used less because of its prolonged action time (5- to 7-day half-life) and increased toxic side effects. Another important cardiac glycoside, ouabain, is primarily used in cardiac emergencies because of its rapid onset and short duration of action, with maximal effects being obtained 30-60 minutes after intravenous administration.

In recent years, newer, more effective cardiac glycosidal drugs have been substituted for many of these older substances. Digoxin, for instance, is often exchanged with verapamil (Isoptin), metoprolol (Lopressor), ditiazem (Cardizem CD), or atenolo (Tenormin) for the treatment of both acute and chronic atrial fibrillation. In contrast, these new drugs are effective within 5 to 10 minutes intravenously, whereas the effects of digoxin occur 3 hours after administration (Rakel, 1996).

## Pharmacology

The principal physiological actions of cardiac glycosides are an increase in the force of myocardial contraction (positive inotropic effect), a decrease in the heart rate, and a reduction in the conductivity of the heart, particularly through the atrioventricular node (Bruneton, 1995). These physiological changes also cause an indirect increase of cardiac output, relieve pulmonary congestion, and alleviate peripheral edema. There is also a direct effect on the vascular smooth muscle and an indirect stimulatory effect on the autonomic nervous system.

The positive inotropic effect of digitalis is apparently due to an increase in intracellular sodium brought on by the glycoside's activity in inhibiting $Na^+, K^+$-activated adenosine triphosphotase (Na/K ATPase). Digitalis also significantly alters the electrical activity of certain cardiac fibers, an influence that is associated with both the therapeutic and toxic effects of these drugs (Gilman et al., 1985). It has also been postulated that the action of the cardiac glycosides may be based on counteracting elevated sympathetic neuronal activity (van Zwieten, 1994).

It is interesting to note that our bodies produce certain of these glycosides. Ouabain, for instance, is secreted as a hormone by the adrenal cortex and plays a part in the regulation of intracellular sodium and in the balance of body salt and water. Ouabain may also be a paracrine hormone, secreted by some central nervous system neurons as well as by other types of cells. Endogenous ouabain, like that extracted from *Strophanus*, directly inhibits the plasmalemmal Na/K-ATPase pump in a variety of cell types. Researchers have also discovered endogenous forms of digoxin, digitoxin, and other cardiac glycosides in the body (Blaustein, 1993).

***Herbs*** listed in the *Botanical Safety Handbook* that contain cardiac glycosides:

| | |
|---|---|
| *Apocynum androsaemifolium* | *Convallaria majalis* |
| *Apocynum cannabinum* | *Digitalis purpurea* |
| *Asclepias tuberosa* | |

## Cyanogenic Glycosides

### Abstract

Cyanogenic glycosides are sugar-containing compounds with a cyano side group. There are 26 cyanogenic glycosides known to occur in plants, all of which are β-glycosidic derivatives of α-hydroxynitriles. These compounds are recognized in over 2100 species of higher plants and are distributed throughout 110 families. The most noted of these are the Rosaceae (containing 150 species), Araceae, Asteraceae, Euphorbiaceae, Fabaceae, and Passifloraceae (Keeler & Tu, 1983).

The best-known cyanogenic glycoside is amygdalin, found in the seeds of domesticated rosaceous fruits, such as cherries, apples, peaches, apricots, and pears. Amygdalin is also found in bitter almonds, causing the familiar acrid smell and taste. The concentration of amygdalin ranges from absent in sweet almonds, to 8.5% in bitter almonds (Keeler & Tu, 1983).

### Adverse Effects

Of the plants included in this survey that contain this class of glycosides, the seeds of several species of *Prunus* present the most serious concern. Apricot, peach, and bitter almond pits can contain up to 8.0% amygdalin, a sufficient concentration to prove fatal following even moderate consumption *See* specific concentration levels at the individual listings. In the interest of caution, long-term use is discouraged for those parts of these plants which are high in cyanogenic glycosides.

Hydrocyanic acid (HCN) is released during the metabolism of these glycosides. HCN disrupts the oxygen-utilizing machinery in the cell's mitochondria. By tightly binding to a protein in the mitochondrial electron transport chain, it instantly prevents the utilization of oxygen and production of adenosine triphosphate, essentially suffocating the body's cells and causing widespread cell death.

The lethal dosage of HCN is approximately 150 mg. Death occurs rapidly, within 12-20 minutes of exposure. Symptoms of acute cyanide poisoning are hyperventilation, headache, nausea, and vomiting, followed by collapse, coma, convulsion, and respiratory failure. Although there are effective antidotes for acute cyanide poisoning, treatment must be given immediately after exposure.

The human body is able to render low amounts of HCN harmless by rapid acting detoxification mechanisms. The enzyme rhodanase can convert 30-60 mg of HCN per hour into relatively nontoxic thiocynate. This compound can then be exhaled or expelled in urine. Nonlethal amounts of HCN are considered teratogenic, however, and lower amounts may cause a severe toxic reaction, displaying symptoms of headache, nausea, and vomiting.

One of the editors of this text (CH) has observed feelings of dizziness, light-headedness, and nausea following several deep inhalations of the crushed leaves of *Prunus virginiana* var. *demissa* (western choke-cherry).

## Therapeutic Use

Plants containing cyanogenic glycosides have a long history of use both as foods and as medicinal substances. Wild cherry bark (*Prunus serotina*) was included in the first edition of the *United States Pharmacopeia* for use as a cough remedy and maintained its listing there, along with the seed of bitter almonds (*Prunus amygdalus* = *P. dulcis*), throughout the 19th century. Loquat leaves (*Eriobotrya japonica*) have long been used for the treatment of coughs, and the use of peach tree leaves (*Prunus persica*) for relief of nausea is equally longstanding. Cassava, a starch derived from species of *Manihot*, is consumed on a daily basis by over 300 million people in tropical countries. Although the food in its raw form contains toxic quantities of manihotoxin, primitive means to remove the toxic constituent were developed centuries ago.

In the 1970s, the use of laetrile, a product consisting mainly of amygdalin derived from apricot kernels, was popular for the treatment of cancer. Rigorous trials, however, found no demonstrable antitumor activity, and, in fact, cyanide poisoning associated with its oral consumption has been reported (Reynolds, 1989).

## Pharmacology

Cyanogenic glycoside-containing plants do not contain detectable free hydrocyanic acid. Cyanide is released when the plant tissue is crushed, chewed, or otherwise disturbed, thereby allowing for the catabolism of the glycosides.

Catabolism is augmented by the enzyme β-glycosidase. Many of the plants that contain these glycosides also contain endogenous levels of this enzyme, resulting in a potential increase of cyanogenic glycoside metabolism (Keeler & Tu, 1983). There is evidence, however, that β-glycosidase is unable to function at a pH below 5 or 6. On the other hand, it has been suggested that the acidic environment in the stomach contributes to a breakdown of cyanogenic glycosides and the release of free cyanide. At this time it is not known with certainty whether cyanognic glycosides require enzymatic action or acid hydrolysis to isolate the contained cyanide (Keeler & Tu, 1983).

There are also studies that show that microfloral enzymes are able to metabolize cyanogenic glycosides. When microflora *Enterobacteria* and *Enterococci* are incubated with amygdalin, the release of free HCN is recorded. It appears that direct contact of cyanogenic glycosides with the gastrointestinal flora maximizes the possibility of cyanide release and subsequent toxicity.

**Herbs** listed in the *Botanical Safety Handbook* that contain cyanogenic glycosides:

*Eriobotrya japonica*
*Hydrangea arborescens*
*Prunus armeniaca*
*Prunus dulcis*
*Prunus persica*
*Prunus serotina*
*Prunus spinosa*

## *Estragole*

### Abstract

Estragole is an allylbenzene with weak hepatocarcinogenic potential. Its chemical structure is 1-allyl-4-methoxybenzene, belonging to the same class of compounds as safrole and β-asarone. Traces of estragole can be found in the essential oils of medicinal plants from the Apiaceae, Lamiaceae, and Asteraceae families.

### Adverse Effects

Estragole is a procarcinogen, becoming carcinogenic following metabolic activation. Animal experiments provide evidence that estragole has weak carcinogenic effects in the liver.

It is reasonable to assume that the carcinogenic risk of estragole is minimal. An extrapolation from published studies on the toxicity of pure estragole would conclude that the estragole-containing herbs should present almost no risk when used periodically at low doses. This conclusion is supported by the observation that rapid detoxification reactions occur immediately after estragole is metabolized into its carcinogenic form. Equally important in determining estragole's carcinogenicity in humans is recognizing that the dose levels of estragole used in animal carcinogenicity studies are several hundred times greater than the estimated human daily intake (Anthony et al., 1987).

For instance, an 8 oz cup of fennel tea contains no more than 0.8 mg of estragole, almost none of which is activated into the carcinogenic form (Fehr, 1982). Nevertheless, until further studies have proven that estragole-containing herbs pose no carcinogenic risk, prolonged use of therapeutic quantities is not recommended.

There is some controversy regarding the safety of the use of estragole-containing herbs during pregnancy and lactation. Fennel has been widely used to promote lactation with no account of toxicity for the mother or infant. *Adverse Effects of Herbal Drugs 1* reports that fennel (*Foeniculum vulgare*) does not represent any special risk in pregnancy and lactation, while the second volume by the same editor warns against the use of basil (*Ocimum basilicum*) at these times (De Smet, 1992; De Smet, 1993).

## Therapeutic Use

The three estragole-containing medicinal herbs reviewed in the *Botanical Safety Handbook* have a long history of use. Basil, a culinary herb, has been used in medicines as an antiseptic and an aperient, that is, as a gentle stimulant to the digestion. Fennel was cultivated by ancient cultures as a food source and utilized for its medicinal properties as a carminative and galactagogue. Finally, tarragon (*Artemisia dracunculus*) is an aromatic culinary herb.

## Pharmacology

Estragole is a procarcinogen that is not directly hepatotoxic or hepatocarcinogenic. It requires enzymatic activation in the liver before achieving full toxicity. Hepatocytic cytochromes are responsible for secreting hydrolyzing enzymes that convert estragole into dihydrodiol and glutathione conjugates and allylic epoxide intermediates which have the potential to generate extensive genetic and cellular toxicity. However, additional enzymes secreted by the liver (epoxide hydrolases and glutathione S-transferases) act to detoxify these epoxides (Luo & Guenthner, 1995). These enzymes quickly inactivate the carcinogenic metabolites of estragole, rescuing the liver from genetic destruction. The inactivation and discharge of estragole's carcinogenic metabolites are extremely efficient. In one human study, researchers determined that the major routes of elimination were in the urine and the expired air. Only 0.2-0.4% of the total estragole dose remained in its carcinogenic form upon elimination (Sangster et al., 1987).

It is interesting to note that the pathway of estragole's activation into its carcinogenic metabolites is dose-dependent. A rat study determined that low doses of estragole undergo O-demethylation and side-chain cleavage resulting in the production of noncarcinogenic metabolites. High doses of estragole, on the other hand, undergo side-chain oxidation that results in the production of carcinogenic metabolites such as the potent carcinogen 1'hydroxyestragole (De Smet, 1992). In contrast, a similar study measuring metabolic effects in humans demonstrated that there are no dose-dependent differences in estragole metabolism with doses ranging from 1 to 250 mg/day (Caldwell & Sutton, 1988).

In a study unrelated to estragole's carcinogenic effects it was determined that the constituent depresses muscle activity in rats and toads. Researchers believe estragole may have two sites and mechanisms of action on muscle fiber, the post-junctional membrane, where neuromuscular transmissions are blocked, and the sacroplasmic reticulum, where myoplasmic calcium is increased (Albuquerque et al., 1995).

**Herbs** listed in the *Botanical Safety Handbook* that contain estragole:

*Artemisia dracunculus*          *Ocimum basilicum*
*Foeniculum vulgare*

## Iodine

### Abstract

Iodine is an essential trace element required for normal thyroid activity. The total amount of iodine in the human body ranges from 20-50 mg, 60-80% of which is found in the thyroid. It is used therapeutically in the treatment of iodine deficiency disorders, such as goiter and hypothyroidism, and administered prophylactically to prevent endemic goiter disease in regions of the world where the diet is deficient in iodine (Reynolds, 1993).

The only abundant, naturally occurring sources of iodine are seaweed and sea vegetables. Land vegetables have a very minute amount of iodine in comparison.

### Adverse Effects

A deficiency or an excess of iodine can lead to a variety of unhealthy conditions, especially of the thyroid gland. Long-term excessive consumption of iodine can result in goiter, caused by the disruption of iodine utilization by the thyroid gland. Symptoms of long-term exposure include irritation of the eyes, severe skin eruptions, gastrointestinal upset, diarrhea, increased salivation, and inflammation of the mouth and throat. Persons with a history of thyroid disease are advised to limit consumption of seaweed. Excessive consumption in pregnancy is also to be avoided, as infantile goiter has been reported.

## Therapeutic Use

Iodine was the first nutrient to be recognized as essential for animal development. As early as 3000 B.C, the Chinese treated goiter by feeding individuals with seaweed and burnt sponge. The Greek physician Hippocrates (460 to 370 B.C.) also used seaweed in the treatment of enlarged thyroid glands (Wichtl, 1994).

Iodine and iodine-containing compounds have an extensive history of use in modern medicine. Although it has now been replaced by pharmaceutical alternatives, bladderwrack (*Fucus vesiculosis*) was once used in therapy for thyroid deficiency (Solis-Cohen & Githens, 1928). Isotopes of iodine, $I^{123}$, $I^{125}$, and $I^{131}$, are used in radiopharmaceuticals for the treatment of thyrotoxicosis and thyroiditis (Reynolds, 1993).

Iodine-based drugs are active against certain bacteria, fungi, viruses, protozoa, cysts, and spores and are used as disinfectants. An iodine solution is often applied to small wounds or abrasions to prevent infection. Iodine tablets are added to water supplies of questionable cleanliness for the purpose of sterilization (Reynolds, 1993).

Dried kelp contains the highest natural concentration of iodine, at 624 mcg per gram. Iodized salt, by comparison, contains 7.6 mcg of iodine per gram. The normal daily requirement for iodine ranges from 100 to 300 micrograms and daily intake of up to one milligram is considered safe. Consumption in excess of this amount may give rise to hyperthyroid conditions.

## Pharmacology

Dietary iodine, in its ionic form of iodide, plays an essential role in the synthesis of thyroid hormones. The conversion of stored thyroglobulin to thyroxine and triiodothyronine is accomplished only after the protein is iodinated in the thyroid gland.

**Herbs** listed in the *Botanical Safety Handbook* that contain iodine:

*Fucus vesiculosus*      *Rhodymenia palmetta*
*Nereocystis luetkeana*

---

# Lectins

## Abstract

Lectins are glycoproteins found in the seeds of the common bean and other plants, primarily in the Fabaceae and Euphorbiaceae families. A class of lectins, known as phytohaemagglutinins, has carbohydrate-binding properties. These compounds have the ability to force red blood cells to bind together, or agglutinate. Many lectins have immune stimulating and anti-cancer properties; some present toxicity concerns (Keeler & Tu, 1983; Reynolds, 1993).

Certain toxic lectins are concentrated in mistletoe leaves (*Viscum album*), all parts of poke weed (*Phytolacca americana*), and in the seeds of castor bean (*Ricinus communis*). Although ricin, the lectin found in castor seeds, is one of the most toxic substances known, it is completely removed during the processing of castor oil.

## Adverse Effects

Medicinal plants containing toxic lectins display a diverse array of side effects. Mistletoe, for example, contains lectins that are highly toxic to the red blood cells, as well as phoratoxin, a small basic protein that is toxic to the heart. Kidney bean and peanut lectins have mild agglutination effects on red blood cells. Soybean lectin is a protease inhibitor (Harborne & Baxter, 1993).

Slightly toxic lectins are also present in green beans and lentils, where the raw seeds and pods contain lectins that result in gastric disturbances. This may be alleviated by cooking and processing the legumes, which denatures the toxic lectins (Bruneton, 1995).

The consumption of highly toxic lectins, such as ricin from castor beans, will often produce a reaction within 2-3 hours, displaying vomiting and bloody diarrhea, loss of fluids, changes in cardiac activity, liver necrosis, shock, and loss of consciousness. The ingestion of only two to four castor seeds by humans can cause nausea, muscle spasms, and purgation. Consumption of as few as eight seeds may result in permanent organ damage and death (Bruneton, 1995; Keeler & Tu, 1983).

## Therapeutic Use

Medicinal plants containing lectins have been used in the folk medicines of India, Egypt, and China as cathartics and applied externally to treat abscesses and sores. In western herbal medicine, pokeweed has similar uses as well as applications as a laxative and for the relief of rheumatic and arthritic inflammation (Lust, 1974). Recent therapeutic utility in the treatment of breast cancer has been reported. Moderate dosage should not be exceeded, as pokeweed is mitogenic, stimulating rapid and uncontrollable cell growth. However, it is safe to use pokeweed in homeopathic remedies for treating rheumatism and arthritis. Mistletoe is used for treatment of hypertension and prevention of arteriosclerosis (Harborne & Baxter, 1993).

In modern medicine, lectin-containing agents are used in a variety of pharmaceuticals and diagnostic procedures, predominantly in the immunological, oncological, and neurological specialties. Phytohaemagglutinin, for example, is used to stimulate interferon-like substances for the measurement and detection of immunocompetence and malignant diseases. Eurixor, a mistletoe extract standardized to the lectin component, is a valuable addition to standard chemotherapy for breast cancer (Werbach & Murray, 1994). Iscador, plenosol, and helixor are other lectin-containing drugs used in the treatment of acute lymphoblastic leukemia and cancer, especially in Germany. In the laboratory, lectins are used in blood work and immunological research. Agglutination reactions are used to detect blood groups and for the separation of leukocytes from erythrocytes. Since lectins are able to bind cell surface glycoproteins, they are also used in tissue biopsy for the differentiation of malignant cells. Diagnostic lectins are also used to distinguish between various strains of infectious organisms.

## Pharmacology

The chemical structure and activity of lectins are highly diversified. A number of lectins have been shown to modulate the human immune system and direct cellular differentiation. Some of these compounds are toxic and may produce adverse effects.

Many lectins, including those found in soybeans, castor, and mistletoe, consist of two pairs of glycoprotein chains. The toxic compound ricin consists of at least four lectins, two agglutinins (RCL I and II), and two toxins (RCL III and IV) that interfere with protein synthesis of the intestinal wall. Absorption of these compounds is sufficiently slow to create a period of latency between consumption and onset of symptoms. Ricin has also been found to have antitumor activity and is used in experimental cancer research; however, the pharmacological pathway is not understood (Fuller & McClintock, 1986; Harborne & Baxter, 1993).

**Herbs** listed in the *Botanical Safety Handbook* that contain lectins:

*Phytolacca americana*       *Viscum album*
*Ricinus communis*

# Oxalates

## Abstract

Oxalates are salts of oxalic acid, a simple dicarboxylic acid with the structure HOOC–COOH. They are widespread in plants, especially from the Oxalidaceae, Araceae, and Polygonaceae families. They occur in the form of insoluble calcium and soluble potassium salts. The compounds may appear in the form of colorless crystals, with considerable volatility and a strong, sour taste. The free acids may also occur.

## Adverse Effects

Oxalic acid combines with calcium in the bloodstream, forming insoluble calcium oxalate and potentially depleting available calcium to deficiency levels. When calcium oxalate crystals are deposited in the kidneys, mechanical injury can occur, leading to renal problems. It is therefore most prudent for those

APPENDIX 1: HERBAL CONSTITUENT PROFILES                                    149

with a history of kidney disease to refrain from the consumption of the herbs listed here.

Pure oxalic acid is an acidic irritant and causes paralysis of the nervous system when taken in soluble form. These concerns for the pure substance are not relevant to the current study.

## Therapeutic Use

The herbs listed in this category are grouped here only because of their common content of oxalates and have no common therapeutic use.

## Pharmacology

Sodium oxalate, when taken internally, causes the precipitation of calcium into calcium oxalate in the blood. This can produce hypocalcemia, a situation which affects the osmoregulation and buffering capacity of the blood. The greatest concern, however, is acute kidney failure by the precipitation of calcium in the renal tubules.

Small amounts of oxalic acid shorten blood coagulation time. It is clinically used for reducing blood clotting time in hemophiliacs and to control bleeding in surgical procedures. The mechanism of action is unknown (Osol & Farrar, 1955).

*Herbs* listed in the *Botanical Safety Handbook* that contain oxalates:

| | |
|---|---|
| *Capsella bursa-pastoris* | *Rumex acetosa* |
| *Portulaca oleracea* | *Rumex acetosella* |
| *Quillaja saponaria* | *Rumex crispus* |
| *Rheum officinale* | *Rumex obtusifolius* |
| *Rheum palmatum* | *Symplocarpus foetidus* |
| *Rheum tanguticum* | *Viburnum prunifolium* |

# *Pyrrolizidine Alkaloids*

## Abstract

Pyrrolizidine alkaloids (PAs) are predominately esters of amino alcohols and necic acids found in some plants of the Asteraceae, Boraginaceae, Fabaceae, and other families. In a study published in 1988, the World Health Organization listed 60 plants containing PAs which are used as medicinal substances. Consumption of certain of the PAs are cumulative poisons and have been associated with fatal liver disease both in humans and, more often, in range livestock. Some specific PAs have also been demonstrated to be mutagenic and carcinogenic.

Not all PAs are toxic. Neither is the presence of a PA in an herb in and of itself a toxicity concern. There are significant variables in chemical structure which are used to differentiate between toxic and harmless PAs. Those with saturated necine bases, such as are found in eyebright (*Euphrasia* spp.) and *Echinacea*, are nontoxic. Toxic PAs, contained in various concentrations in the plants enumerated in this section, are those with an unsaturated necine ring. All of these are reported to include a 1,2-double bond, esterified hydroxy group, and at least one branched carbon side chain (Bruneton, 1995; De Smet, 1992).

## Adverse Effects

Reports of acute poisonings of livestock from consumption of *Senecio* and *Amsinckia* are not uncommon (Fuller & McClintock, 1986; Turner & Szczawinski, 1991). Acute concerns in humans are primarily associated with accidental adulteration of food supplies. The primary emphasis of this examination, however, is on the potential for serious liver damage related to chronic consumption of the species listed here which have been established in a therapeutic framework.

The herbs most widely used in the United States which present risks due to the presence of PAs are comfrey root and leaf (*Symphytum* spp.), coltsfoot leaf and flower (*Tussilago farfara*), and borage leaf (*Borago officinale*), as well as several species of *Eupatorium*. Relative toxicities vary widely depending on the part of the plant consumed and from one species to another of the same genus. For example, the concentration of alkaloids in comfrey is measured at about 10 times higher in the root than in the leaf (Tyler, 1994). Moreover, echimidine, the most toxic of the alkaloids found in comfrey, is present in the so-called Russian comfrey species (*S. asperum* and its cultivars) but is completely absent in most cytotypes of *S. officinale* (Awang et al., 1993; Tyler, 1994). Except for those manufacturers of botanical supplements who use botanically referenced materials or analyze for echimidine, the toxic species is not readily excluded from trade.

Although some of these alkaloids have shown carcinogenic and mutagenic properties, and though renal toxicity is reported, the primary concern for use of these herbs is the potential for serious liver damage, specifically hepatic veno-occlusive disease. This potentially fatal condition manifests symptoms such as abdominal pain, diarrhea, vomiting, swelling of the liver and spleen, accumulation of fluid in the abdominal cavity, jaundice, cirrhosis of the liver, and liver failure.

Cautious restrictions on the use of all of the herbs containing unsaturated (toxic) PAs have been recommended by AHPA, with suggestions to limit use to external application on unbroken skin only and to refrain from use while nursing. All use is contraindicated in pregnancy and in persons with a history of liver disease. Use in Australia is confined to homeopathic preparations, and Canadian regulations do not permit these herbs as ingredients in foods

APPENDIX 1: HERBAL CONSTITUENT PROFILES 151

(Baker, 1990; Blackburn, 1993). Quantified limits are established in Germany for some of these herbs, the details of which are included at each of the separate entries in the body of this text.

## Therapeutic Use

The historical internal use of these herbs has been significantly challenged by the realization of potential toxicity associated with long-term use and in persons with compromised hepatic health. In most cases, other herbs should be recommended as substitutions for these for their established uses.

Both the leaves and root of comfrey contain allantoin, a naturally occurring substance which promotes tissue growth (Osol & Farrar, 1955; Wichtl, 1994). The plant's high concentration of mucilage provides a rationale for comfrey's historical usage in treating stomach ulcers, inflammatory bowel disease, and broken bones (Wren, 1988). The leaves are also used as a soothing demulcent for upper respiratory conditions, especially in cases where the mucosal lining is dry and irritated. Comfrey salves continues to be widely used and present little toxicological concern, though these should not be used for breast tenderness during lactation.

The therapeutic uses of other PA-containing herbs have been long established: borage leaves for pulmonary complaints and as a diuretic; gravel root (*Eupatorium purpureum*) as a treatment for urinary calculi; coltsfoot (*Tussilago farfara*) for upper respiratory congestion (Felter & Lloyd, 1898). This last herb is sometimes adulterated with western coltsfoot (*Petasites* spp.), an herb with a considerably higher concentration of toxic PAs.

Life root (*Senecio aureus*) has historically been used as a diuretic, emmenagogue, and uterine tonic (Wren, 1988). Some contemporary herbal texts continue to prescribe the use of this herb as a uterine tonic, a practice that is not prudent due to the particular susceptibility of the early stage fetus and the lapse in time between conception and knowledge of pregnancy.

## Pharmacology

Hepatotoxicity associated with chronic consumption of unsaturated PAs is apparently due to enzymatic conversion of the alkaloids to toxic pyrroles, esters which act as alkylating agents and which are destructive to the liver's tissue (Bruneton, 1995).

**Herbs** listed in the *Botanical Safety Handbook* that contain pyrrolizidine alkaloids:

*Alkanna tinctoria*  
*Borago officinalis*  
*Eupatorium purpureum*  
*Symphytum asperum*  
*Symphytum* x *uplandicum*  
*Symphytum officinale*  
*Tussilago farfara*

## Safrole

### Abstract
Safrole is a minor component of the aromatic oils of nutmeg (*Myristica fragrans*), cinnamon (*Cinnamomum verum*), and camphor (*Cinnamomum camphora*), and a major constituent of oil of sassafras. Like estragole and β-asarone, safrole is an allylbenzene with weak hepatotoxic, carcinogenic, and mutagenic potential. Its chemical structure is 1-allyl-3,4-methylenedioxybenzene, containing one aromatic ring, an intact allyl, and a methylenedioxyphenyl group. Traces of safrole are found throughout the plant kingdom, recorded in 53 species and varieties, representing 10 families (Keller & Tu, 1983).

### Adverse Effects
Certain safrole-containing herbs, including cinnamon, camphor, and nutmeg, should be avoided during pregnancy except in minor amounts as food flavorings. A study measuring transplacental exposure to safrole in mice determined that safrole readily crosses the placenta, resulting in DNA damage within the rapidly proliferating tissues of the fetus (Lu et al., 1986). Although De Smet reports that cinnamon does not present any special risk in pregnancy, he recommends that prolonged use of the oil be restricted (De Smet, 1992), a caution that is also relevant to camphor.

Safrole is a weak carcinogen in laboratory animals when taken in high doses and for extended periods of time (Keeler & Tu, 1983; Wrba et al., 1992). It has also demonstrated hepatocarcinogenicity in rodents (Chan & Caldwell, 1992), an action related to its ability to induce unscheduled DNA synthesis in freshly isolated rat hepatocytes in primary culture (Howes et al., 1990). Mice given safrole-containing beverages for 8 weeks developed significant levels of covalent liver DNA adducts leading to hepatoma. These adducts were not detected in mice given water or non-safrole beverages. This same team of researchers found identical DNA adducts when mice were treated with extracts of nutmeg (Randerath, 1993).

Like the other allylbenzenes, safrole is a procarcinogen, becoming carcinogenic following metabolic activation. Safrole has reportedly failed to demonstrate mutagenicity in fruit flies (Batiste-Alentorn et al., 1995) or mutagenicity or carcinogenicity in bacterial assays (Carls & Schiestl, 1994), including the Ames test (Howes et al., 1990). An *in vitro* micronucleus test performed on human lymphocytes determined that safrole was non-genotoxic (Howes et al., 1990).

There is no direct evidence that consumption of safrole-containing herbs has a carcinogenic activity in humans, though *in vitro* studies using human DNA and liver enzymes have demonstrated potential carcinogenicity with safrole. Regulatory restrictions have been placed on sassafras and/or safrole by the U.S. FDA and by the state of California, although all of these herbs continue to be readily accessible (Heikes, 1994; Keeler & Tu, 1983). While conclusive work is lacking, it is reasonable to assume that safrole might induce the same effects in humans as have been shown in laboratory animals. Further work is necessary to fully evaluate the safety of these important natural flavors.

## Therapeutic Use

Sassafras is a prolific tree indigenous to the eastern U.S. It was the first product exported to England by the early European settlers in the 17th century, who learned of its medicinal uses from the native Americans. It is reported to have diaphoretic and diuretic actions and was used in the treatment of colds, flu, and arthritic pains, as well as for skin diseases like acne. An extract of sassafras was used for many years as a flavoring agent in the soft-drink industry, providing one of the familiar natural root beer flavors. In 1960, however, researchers began to question the safety of safrole, and the use of sassafras in food products was subsequently limited to desafrolized products. Contemporary use of sassafras oil is primarily limited to the manufacture of perfumery and soaps (Grieve, 1931).

Camphor has rubefacient, parasiticide, antiseptic, analgesic, and diaphoretic actions. Most herbal preparations containing camphor are for external application and include composite oils to relieve pain, muscle tension, arthritis, rheumatism, and respiratory congestion. Nutmeg, a common spice, has analgesic, stomachic, and carminative actions and is sometimes recommended in the treatment of diarrhea, indigestion, colic, and flatulence. Cinnamon has diaphoretic, astringent, and stomachic actions and is used for treating colds, flu, and fevers in traditional Chinese and Ayurvedic medicine and other traditional systems of healing. It is a popular flavoring agent, added to commercial food products, teas, and products such as toothpaste and mouthwash.

## Pharmacology

Safrole, like estragole and β-asarone, though not directly hepatoxic or hepatocarcinogenic, is readily converted by enzymatic activation in the liver into substances which are known to be toxic. Hepatocytic cytochromes (P-448 and P-450) secrete hydrolyzing enzymes that convert safrole into dihydrodiol and glutathione conjugates and allylic epoxide intermediates which have the potential to generate extensive genetic and cellular toxicity. The most potent safrole intermediate is 1'-hydroxy-2',3'-dehydroxysafrole (Wiseman et al., 1987). A number of additional bioactivation pathways also contribute to safrole's carcinogenic effect (Iverson et al., 1995; Miller & Miller, 1983). However, simultaneous enzymatic activities occur that act in a protective manner to detoxify these epoxides (Luo & Guenthner, 1995).

The safrole metabolites that are not deactivated by detoxification enzymes have the potential to achieve carcinogenesis. Safrole's carcinogenic metabolites are free radicals and strong electrophiles that are able to bind covalently and non-enzymatically with nucleophilic sites in DNA and RNA proteins and small molecules in target tissues (Miller & Miller, 1983). The formation of covalent adducts between DNA and safrole's reactive intermediates, for example, results in mutagenesis. A number of studies have demonstrated safrole's carcinogenic activity in animals (Gupta et al., 1993; Qato & Guenthner, 1995), as well as in humans (Ireland et al., 1988).

**Herbs** listed in the *Botanical Safety Handbook* that contain safrole:

*Cinnamomum camphora*        *Ocimum basilicum*
*Cinnamomum verum*           *Sassafras albidum*
*Myristica fragrans*

# Salicylates

## Abstract

The salicylates are salts or esters of salicylic acid, a hydroxybenzoic acid derived from naturally occurring or synthetic salicin. Salicin is found in the bark, leaves, and flowers of most species of willow (*Salix*) as well as in each of the other plants listed here. Salicin was first proposed as the active ingredient of willow bark in 1828 by Buchner (Mayer & Mayer, 1949). Salicylic acid was synthesized in 1860 and the salicin-containing plants were soon supplanted by the synthetic analog, acetylsalicylic acid, better known as aspirin (Weissman, 1991). To this day, the salicylates are the most widely consumed analgesics, especially for low intensity pain (Gilman et al., 1985).

## Adverse Effects

Concern for the consumption of the salicin-containing plants is addressed here primarily to assure that the known adverse effects of aspirin are examined in relationship to this naturally occurring related compound. While persons with known sensitivity to aspirin and other salicylates may wish to exercise caution with these plants, there is no evidence that the types of reactions known to be associated with the pharmaceutical salicylates is observed with *Salix* or any other of these species. Likewise, *in vitro* studies have demonstrated that salicin's physiological actions are sufficiently different from those of acetylsalicylic acid, leading to some assurance that the tendency seen in aspirin to potentiate blood thinning drugs is not mimicked by the plants' constituents (Wichtl, 1994). Side effects commonly associated with aspirin, such as allergic reactions and Reye's syndrome, have not been observed with salicin-rich plants.

## Therapeutic Use

Salicin-containing plants have been used continually for centuries. Willow was described by Dioscorides in the first century, listed in the first edition of the *Pharmacopeia of the United States* in 1820, and continues to be used by herbalists today. Its primary uses are for the treatment of fever and as a mild analgesic and anti-inflammatory.

The leaves of wintergreen (*Gaultheria procumbens*) and the bark of the sweet birch (*Betula lenta*) are sources of essential oils made up almost completely of methyl salicylate. These oils have been used extensively to provide natural root beer flavor. Historical uses of the plants are similar, in that both have been used for fever and stomachache (Leung & Foster, 1996).

## Pharmacology

Most research has focused on the ability of salicylates to suppress the synthesis of prostaglandins, hormones thought to play an integral role in pain, inflammation, and fever. Two specific enzymes, cyclooxygenase 1 and 2 (COX1 and COX2), are considered to be predominant in this process. COX1 occurs in platelets, blood vessels, and other organs; COX2 acts primarily in inflamed tissue.

Aspirin is the most commonly used salicylate. It blocks the synthesis of prostaglandins through the acetylation of cyclooxygenase, especially COX1, by an irreversible transfer of the acetyl group into the enzyme. Salicylic acid and salicylates (such as salicin) that lack an acetyl group are not as effective as aspirin in inhibiting platelet aggregation. Therefore, there is little concern for salicin-containing plants causing haemotological disturbances. Conversely, these plants are not appropriate as a preventative treatment against stroke, a benefit associated with aspirin/calcium consumption.

**Herbs** listed in the *Botanical Safety Handbook* that contain salicylates:

| | |
|---|---|
| *Betula lenta* | *Gaultheria procumbens* |
| *Betula pendula* | *Populus balsamifera* |
| *Betula pubescens* | *Populus x jackii* |
| *Filipendula ulmaria* | *Salix alba* |

# Tannins

## Abstract

Tannins are a broad class of complex phenolic compounds. The biological importance of tannins is attributed to their ability to bind with and precipitate proteins. They are natural components of many herbs and common foods and, in some forms, are used in the processing of foods, alcoholic beverages, and medicines. The most highly concentrated sources of tannins are in the oak family (Fagaceae), especially in oak galls, which contain up to 80% tannins (Evans, 1989; Leung & Foster, 1996).

Tannins are classified into two groups: the hydrolyzable tannins, which, upon acidic, alkaline, or enzymatic hydrolysis produce glucose and phenolic acids; and the condensed tannins (proanthocyanidins), which are flavan-3-ol polymers. Both types of tannins are found in certain plants, including witch hazel (*Hamamelis virginiana*) and tea (*Camellia sinensis*). All tannins have astringent properties, providing the basis for many of the historical medicinal uses of the plants containing them.

## Adverse Effects

Tannins are broadly distributed throughout the plant kingdom, occurring in the barks, roots, leaves, fruits, seeds, and other parts of many different species. Only those plants which are reported to contain at least 10% tannins have been identified as relevant to this discussion of the potential adverse effects of tannin consumption.

Most of the known adverse effects related to tannins are specifically recorded for consumption of tannic acid, an ethereal or hydroalcoholic extract of nutgalls (from *Quercus* spp.), and include gastrointestinal disturbances and kidney damage, as well as severe necrotic conditions in the liver (Gilman et al., 1985; Osol & Farrar, 1955; Reynolds, 1989). While all of these concerns may be relevant to the use of high tannin content herbs, only the digestive irritating properties of tannins are traditionally associated with the consumption of these.

Both carcinogenic and anti-cancer properties have been reported in experimental settings that measured the effect of tannins on laboratory animals (Leung & Foster, 1996). A correlation has been made between increased esophogeal or nasal cancer in humans and regular consumption of certain herbs with high tannin concentrations (Lewis & Elvin-Lewis, 1977). An analysis of medical records from the 19th century documents an increase in esophageal cancer in Britain with the introduction of tea (Keeler & Tu, 1983). It has been proposed that any adverse effect of tannins in tea might be mitigated by the addition of milk to tea, which effectively binds the tannins to a protein in the milk.

## Therapeutic Use

Tannic acid has been used as an astringent for treating diarrhea, spongy or receding gums, sore throat, inflamed tonsils and hemorrhoids, and also as a styptic. Its historical external use for the treatment of burns is no longer recommended due to associative liver damage, though such concern may in fact be due to impurities in the pharmaceutical form. External use is also established for skin conditions, including acute dermatitis and poison ivy irritation (Leung & Foster, 1996; Osol & Farrar, 1955; Reynolds, 1989).

The plants listed in this category enjoy a wide range of therapeutic uses, not all of which are associated with their contained tannins. Herbal therapies which are related to tannins include the use of oak bark (*Quercus* spp.) for certain skin conditions, as a gargle or mouthwash for sore throat and gum disease, as a vaginal douche for leukorrhea, and as an infusion for diarrhea. Similar uses are ascribed to geranium (*Geranium maculatum*), rhatany (*Krameria* spp.), and gambir (*Uncaria gambir*) (Solis-Cohen & Githens, 1928; Weiss, 1988).

Recent research on various health benefits of some tannin-containing herbs has focused on the related catechins and procyanidins in tea, pine bark, and grape seeds. These substances are reported to be strong antioxidants and free-radical scavengers. Statistical evaluations of populations with high tea consumption as well as animal studies have shown antimutagenic properties and protective effects against liver and heart disease (Bombardelli & Morazzoni, 1995; Imai & Nakachi, 1995; Leung & Foster, 1996).

## Pharmacology

The therapeutic activities of tannins are associated with their ability to bind with and precipitate proteins and to force dehydration of mucosal tissues. In external use, these actions allow the formation of a protective layer of harder, constricted cells; internally, both normal and pathologic secretions of all types are reduced. Tannins also have the ability to precipitate other macromolecules such as cellulose and pectins through the formation of covalent bonds (Bruneton, 1995; Solis-Cohen & Githens, 1928).

**Herbs** listed in the *Botanical Safety Handbook* that contain over 10% tannins:

*Agrimonia eupatoria*
*Albizia julibrissin*
*Arctostaphylos uva-ursi*
*Camellia sinensis*
*Castanea sativa*
*Eucalyptus globulus*
*Hamamelis virginiana*
*Heuchera micrantha*
*Hypericum perforatum*
*Ilex paraguayensis*
*Krameria argentea*
*Krameria triandra*
*Myrica cerifera*
*Myrica pensylvanica*
*Paullinia cupana*
*Potentilla erecta*
*Punica granatum*
*Quercus alba*

*Quercus robur*
*Quercus petraea*
*Quillaja saponaria*
*Rheum officinale*
*Rheum palmatum*
*Rheum tanguticum*
*Rhus coriaria*
*Rhus glabra*
*Rumex acetosa*
*Rumex acetosella*
*Rumex crispus*
*Rumex hymenosepalus*
*Rumex obtusifolius*
*Salix alba*
*Terminalia bellerica*
*Terminalia chebula*
*Uncaria gambir*

## *Thujone*

### Abstract
Thujone is a bicyclic monoterpene component of numerous volatile oils. The substance is neurotoxic, acting as a convulsant and hallucinogen. Regulations in many countries limit the sale of oral-use products containing certain of these herbs and their essential oils, unless thujone-free. Wormwood (*Artemisia absinthium*) contains approximately 1.5-1.7% of an essential oil rich in thujone and formed the principle ingredient of absinthe, an intoxicating liquor that was banned in the U.S. and most European countries early in the 20th century (Leung & Foster, 1996; Wichtl, 1994).

### Adverse Effects
The adverse effects of thujone consumption are associated with use of the essential oils and alcoholic extracts of plants high in this ketone. Because the essential oils are not efficiently extracted in water, aqueous forms such as tea are low in thujone (Windholz et al., 1983).

Long-term use or high dosage of products high in thujone can cause restlessness, vomiting, vertigo, tremors, renal damage, and convulsions (Reynolds, 1993; Wichtl, 1994). A study demonstrating neurotoxicity in rats found that 0.5 g/kg of sage oil produced convulsions, and death resulted in many of the animals at a dosage of 3.2 g/kg (Millet et al., 1981). The $LD_{50}$ in mice has been measured at 87.5 mg/kg for $\alpha$-thujone, and 442.2 mg/kg for $\beta$-thujone (Windholz et al., 1983).

### Therapeutic Use
The therapeutic uses of the plants that contain thujone are diverse and are not related to the contained terpenes.

### Pharmacology
Thujone is related to camphor in its chemical structure and occurs as two common isomers, $\alpha$-thujone (l-form), and $\beta$-thujone (d-form). There is limited research on its pharmacology. It has been suggested that thujone interacts with the same central nervous system receptor sites as tetrahydrocannibinol, the psychoactive compound in marijuana (Reynolds, 1993). Thujone intoxication exhibits pyschoactivity similar to that noted for cannabinoid use, though thujone does not mimic the cannabinoids in inhibiting the synaptosomal enzyme (Greenberg et al., 1978).

Mammalian cells have been shown to metabolize thujone by reducing the carbonyl group to yield secondary alcohols (Ishida et al., 1989). Thujone causes convulsant action of central nervous system origin (Millet, 1981) and has been demonstrated to have antioxidative effects and moderate antimicrobial and antifungal properties (Graven et al., 1992).

Of additional interest is a study that measured an increase in porphyrin (pigment) production in chick embryo liver cells associated with thujone. Thujone and affiliated terpenes may therefore be hazardous to patients with underlying defects in hepatic heme synthesis. The study also found that thujone induces synthesis of a rate-controlling enzyme for the 5-aminolevulinic acid synthase pathway. Well-known porphyrogenic chemicals such as phenobarbital and glutethimide have similar effects (Bonokovsky et al., 1992).

**Herbs** listed in the *Botanical Safety Handbook* that contain thujone:

*Artemisia absinthium*
*Evernia furfuracea*
*Evernia prunastri*
*Platycladus orientalis*

*Salvia officinalis*
*Tanacetum vulgare*
*Thuja occidentalis*

# Appendix 2

## Herbal Action Profiles

## *Abortifacients*

## Definition
Abortifacients are agents used to terminate pregnancy.

## Adverse Effects
While herbs have been widely employed both as contraceptives and abortifacients, there are little reliable data on their effectiveness, toxic levels, or possible effects upon the developing fetus. Unsupervised consumption of all of the herbs listed in this category is advised against in pregnancy.

Although abortifacients are particularly toxic when taken in large doses or when used over a long period of time, relatively small dosages of some substances, or amounts which an uninformed consumer might believe to be insignificant, have resulted in death. In a fatal event recorded in 1978 following consumption of one ounce of pennyroyal oil, an amount that represents an extremely high dose, the victim suffered two heart attacks, liver and kidney failure, and disseminated vascular coagulation before her death (Chalker & Downer, 1992).

Signs of pennyroyal oil toxicity include nausea, vomiting, sweating, chills, fever, headache, ringing in the ears, dizziness, low blood pressure, difficulty in swallowing, extreme thirst, diarrhea, rapid pulse or heartbeat, muscle spasms, restlessness, drowsiness, unusual talkativeness, fatigue, and tremor. More dramatic symptoms of toxicity include hallucinations, mania, collapsing, convulsions, and coma. Presentation of any of these symptoms requires immediate cessation of use and emergency medical treatment. The preferred agent for emergency treatment of pennyroyal poisoning is N-acetylcycsteine (NAC), an anti-oxidative agent that protects against endotoxemia (Zhang et al., 1994).

Many abortifacients are cytotoxic and teratogenic and are likely to cause severe developmental abnormalities. Prolonged ingestion of certain abortifacient herbs while pregnant may cause nerve damage or other negative effects and fetal abnormalities if the pregnancy is carried to term (Anderson, 1996).

## Therapeutic Use
Inducing abortion with herbal abortifacients is dangerous, since toxic amounts of these herbs are required to force the uterus to expel a healthy fetus. The use of certain of these herbs for other therapies is safe within a much lesser dosage range.

## Mechanism of Action

Abortifacients have many different mechanisms of action. Some abortifacients act indirectly, meaning that they induce abortion through peripheral systems such as the endocrine, cardiovascular, gastrointestinal, or nervous systems. Others are direct acting abortifacients, which target the uterus, endometrium, and/or fetus directly, causing abortion to commence. It is not possible to generalize the action, efficacy, or safety of abortifacients, since each herb has its own unique metabolism and performance. In addition, the mechanism of action has not been well studied for many of the plants listed in the *Botanical Safety Handbook* as abortifacients (Bingel & Farnsworth, 1980).

Pennyroyal (*Mentha pulegium* and *Hedeoma pulegioides*) is an example of a direct acting abortifacient. Pennyroyal's active abortifacient constituents, pulegone and methofuran, are found in its volatile oil, which is believed to have an oxytocic effect, directly stimulating uterine contractions. Pennyroyal is extremely toxic, with as little as 10 ml of the essential oil causing lethal hepatotoxic effects. Protein bound adducts of the metabolites from pennyroyal were found in human livers of women poisoned by pennyroyal (Anderson, 1996).

Another direct acting abortifacient is *Cytisus scoparius*, the abortifacient properties of which are derived from lectins. Lectins are believed to bind and prohibit the functioning of specific galactose-rich glycoproteins that are maximally expressed during pregnancy. These glycoproteins are utilized by the endometrial glands and the luminal epithelium to sustain cellular growth during pregnancy. When these glycoproteins are bound to lectins, they are no longer able to maintain the endometrium. This results in the abortion of the fetus and sloughing off of the endometrial wall (Horvat, 1993).

*Trichosanthes kirilowii* induces abortion by actions associated with certain contained proteins. The abortifacient proteins are trichosanthin and momorcharin, known to inhibit cellular protein synthesis and decrease secretion of human chorionic gonadotrophin (HCG) and progesterone by choriocarcinoma cells. HCG is a hormone required to maintain pregnancy. When it decreases to a critical level, termination of pregnancy is likely (Tsao, 1990). Extracts of tricosanthin have demonstrated potent abortifacient, embryotoxic, and teratogenic activity in pregnant mice. Surviving fetuses had lethal abnormalities including exencephaly and micromelia (Zheng, 1995; Chan, 1993).

Certain abortifacients have drastic purgative effects or are gastrointestinal irritants that can produce reflex uterine contraction. Many volatile oils, such as oil of tansy (*Tanacetum vulgare*), and saponin glycosides, such as those found in beth root (*Trillium* spp.), act in this manner.

# APPENDIX 2: HERBAL ACTION PROFILES

**Herbs** listed in the *Botanical Safety Handbook* with potential abortifacient action:

*Andrographis paniculata*
*Carthamus tinctorius*
*Catharanthus roseus*
*Caulophyllum thalictroides*
*Chamaemelum nobile*
*Crocus sativus*
*Cullen corylifolia*
*Cytisus scoparius*
*Gossypium herbaceum*

*Gossypium hirsutum*
*Rosmarinus officinalis*
*Ruta graveolens*
*Tanacetum vulgare*
*Thuja occidentalis*
*Trichosanthes kirilowii*
*Vetiveria zizanoides*
*Withania somnifera*

## *Bulk-forming Laxatives*

### Definition

Bulk-forming laxatives are substances that promote bowel evacuation by increasing the bulk volume and water content of the stool. These are generally considered to be the safest laxative agents, though see *Adverse Effects* below.

### Adverse effects

Bulk-forming laxatives are contraindicated in bowel obstruction and must be taken with adequate liquid to avoid paradoxical constipation and esophageal or bowel obstruction. FDA requires special labeling of all over-the-counter drug products which contain certain of the herbs listed here, including agar (*Gelidiella acerosa* and *Gelidium* spp.), guar gum (*Cyamopsis tetragonolobus*), and psyllium (*Plantago* spp.) if these are marketed in a "dry or incompletely hydrated form", such as capsules and powders. The designated labeling, which the regulation specifies be printed in bold print and capital letters, is as follows:

"WARNING: TAKING THIS PRODUCT WITHOUT ADEQUATE FLUID MAY CAUSE IT TO SWELL AND BLOCK YOUR THROAT OR ESOPHAGUS AND MAY CAUSE CHOKING. DO NOT TAKE THIS PRODUCT IF YOU HAVE DIFFICULTY IN SWALLOWING. IF YOU EXPERIENCE CHEST PAIN, VOMITING, OR DIFFICULTY IN SWALLOWING OR BREATHING AFTER TAKING THIS PRODUCT, SEEK IMMEDIATE MEDICAL ATTENTION."

Additional language is required under the directions for use, as follows:

"DIRECTIONS: TAKE (OR MIX) THIS PRODUCT (CHILD OR ADULT DOSE) WITH AT LEAST 8 OUNCES (A FULL GLASS) OF

WATER OR OTHER FLUID. TAKING THIS PRODUCT WITHOUT
ENOUGH LIQUID MAY CAUSE CHOKING SEE WARNING."

It is suggested that the above language be posted at the point of sale in any retail setting where any of these products are sold in bulk.

An additional concern associated with these herbs is their ability to inhibit the absorption of other drugs. The drugs usually associated with this consideration are aspirin, digitalis and other cardiac glycosides, antibiotics, and anticoagulants. Active monitoring of the absorption of these drugs, or the avoidance of simultaneous medication is suggested (Zimmerman, 1983). The absorption of dietary nutrients including calcium, iron, zinc, sodium, and potassium can also be inhibited. Appropriate supplementation must therefore be advised when using bulk-forming laxatives for extended periods.

For individuals accustomed to a low-fiber diet, gradual use of less refined fiber in the diet is recommended before using bulk fiber agents.

## Therapeutic use

Because bulk-forming laxatives are high in fiber, they can be used to treat constipation or diarrhea. They are also used to decrease absorption of cholesterol in the blood stream and stabilize blood sugar.

## Mechanism of action

Bulk-forming laxative herbs include gel-forming fibers such as psyllium husk (*Plantago ovata*) and flax seed (*Linum usitatissimum*). Gel-forming fibers contain a form of starch called mucilage, composed of mucopolysaccharides, and roughage or indigestible plant fiber called cellulose. These plant starches are hydrophyllic, absorbing water or other liquid to form a mucilaginous or gel-like substance. Because these herbs also expand on contact with liquid, they add moisture and bulk to stools in the colon. The gel-like starch has the effect of softening feces, making hardened stool easier to eliminate without damage to the bowel wall. Also, since fiber stimulates peristalsis by activating stretch receptor cells in the bowel wall, there is less strain on moving the bowels.

Bulk-forming laxatives that contain mucilage have additional minor benefits complementing their primary effect of relieving constipation. Mucilaginous herbs are demulcent, meaning that they are soothing to inflamed mucosal surfaces. Demulcents form a temporary gelatinous barrier which protects the intestinal wall from irritation caused by caustic material in the intestines, thus allowing repair of the adjoining tissues.

Besides providing the demulcent properties associated with these plants' mucilage content, the indigestible cellulose fiber of bulk-forming laxatives plays additional roles related to diet and digestion. Fiber absorbs dietary fats, decreasing the absorption of cholesterol into the bloodstream. In addition, since dietary fiber from bulk laxatives cannot be digested, these herbs add a feeling of fullness without calories. Fiber also slows the release of dietary sugar from the digestive tract into the blood stream, assisting in the stabilization of blood sugar.

**Herbs** listed in the *Botanical Safety Handbook* as bulk forming laxatives:

| | |
|---|---|
| *Cyamopsis tetragonolobus* | *Gelidium pacificum* |
| *Gelidiella acerosa* | *Gelidium vagum* |
| *Gelidium amansii* | *Linum usitatissimum* |
| *Gelidium cartilagineum* | *Plantago arenaria* |
| *Gelidium crinale* | *Plantago ovata* |
| *Gelidium divaricatum* | *Plantago asiatica* |

## Emetics

### Definition
An emetic is a substance that induces vomiting.

### Adverse Effects
The concerns related to herbs with emetic potential are of primary significance only when consumed in dosage levels sufficient to produce emesis. Contraindications for such usage are pregnancy, high blood pressure, esophageal varices, hiatus hernia, gastritis or peptic ulceration, and recent consumption of central nervous system stimulants. Use of emetics for more than three to four days can produce a serious medical condition if the assimilation of fluids is disrupted. This can lead to dehydration and severe electrolyte imbalances. Continual retching action from chronic emesis will strain the abdominal, gastric, and diaphragm muscles causing severe cramping and potential development of hernias (Rakel, 1996; Hardman & Limbird, 1996; Katcher et al., 1983).

Strong emetics should not be administered to unconscious or deeply sedated individuals or in the event of convulsions, since emesis may cause aspiration of the gastric contents resulting in obstruction of the air passages. The danger of esophageal perforation prohibits use following oral consumption of such caustic substances as strong acids or alkalis (e.g., drain cleaners). The use of ipecac as an emetic following ingestion of petroleum distillates, such as kerosene or petroleum-based household furniture polish, was formerly avoided due to the concern for aspiration of hydrocarbons. Studies have shown, however, that the occurrence of aspiration pneumonitis is both less frequent and less severe following emesis induced by ipecac than that which results from gastric lavage. While medical supervision is strongly suggested, Syrup of Ipecac is at this time the method of choice for treatment of petroleum distillate ingestion.

Activated charcoal should not be administered with emetics, because the charcoal can absorb the emetic substance and reduce the emetic effect. In the event that emesis does not occur following administration of ipecac, gastric lavage should be performed to avoid a toxic reaction to the contained emetine (Rakel, 1996; Katcher et al., 1983; Gilman et al., 1985).

Most poisonings that require emesis also require immediate professional attention. Promptly reporting any such event to an emergency medical facility and/or a poison control center is advised (Hardman & Limbird, 1996).

## Therapeutic Use

The historical record of employing emetic herbs along with purgatives and diaphoretics to cleanse the system and purify the body is extensive. Emetic herbs have also been used to expel bile from the stomach, intestines, and gallbladder, and to aid in the removal of gallstones. Modern phytotherapy occasionally prescribes emetic herbs in very small doses for their action on the gastric mucosa and the related response in the lungs where the effect is as a stimulating expectorant. Some emetic herbs, such as lobelia (*Lobelia inflata*) and ipecac (*Cephaelis ipecacuanha*), have been used in the treatment of chronic, congestive lung conditions, such as pneumonia, bronchitis, or a chronic smoker's cough. For these uses, emetic herbs are generally not recommended in doses sufficient to cause vomiting; and simultaneous administration of carminatives is useful in minimizing associated cramping (Ellingwood, 1983; Solis-Cohen & Githens, 1928).

The primary contemporary use of emetics is to remove ingested poisonous substances prior to their absorption.

## Mechanism of Action

There are two types of emetics. Local emetics act by irritating the end organs of the gastric, pharyngeal, or esophageal nerves. These stimulate the vomiting center in the medulla reflexively. Systemic or general emetics act through the circulatory and/or central nervous system by direct stimulation of chemoreceptors in the medulla and cerebellum.

Emesis is mediated by the vomiting center through various efferent pathways including the vagus, pherenic nerves, and spinal innervation of the abdominal muscles. Vomiting is induced when the upper portion of the stomach relaxes, the pylorus constricts, and the coordinated contraction of the diaphragm and abdominal muscles ejects the gastric contents (Hardman & Limbird, 1996).

APPENDIX 2: HERBAL ACTION PROFILES

**Herbs** listed in the *Botanical Safety Handbook* as emetics:

*Asarum europaeum*
*Asclepias tuberosa*
*Cephaelis ipecacuanha*
*Euphorbia pilulifera*
*Genista tinctoria*

*Iris versicolor*
*Iris virginica*
*Lobelia inflata*
*Melia azedarach*
*Sanguinaria canadensis*

## *Emmenagogues/Uterine Stimulants*

## Definition

Emmenagogues are agents that stimulate and promote menstruation. In the past, the term has also been used for substances intended to restore monthly regularity to the menstrual cycle. Uterine stimulants tonify and strengthen the muscle tone, glandular tissue, and blood supply of the uterus.

Although the emmenagogue herbs are important in modern herbal therapeutics, the category is not well recognized by western medicine, and there is little understanding or research regarding their mechanism of action, therapeutic range, or safety. Much of the following information comes from late-19th century references, such as Biddle's *Materia Medica and Therapeutics* and Felter & Lloyd's *King's American Dispensatory* and the experience of practicing herbalists.

## Adverse Effects

Since many emmenagogues and uterine stimulants have the potential to induce miscarriage, they should be avoided in therapeutic quantities during pregnancy. Other contraindications include pelvic inflammatory disease and excessive bleeding during menstruation (Chalker and Downer, 1992). The use of uterine stimulants to induce labor should always be under the direct supervision of a qualified and experienced individual, usually a midwife versed in such therapy. If used improperly, these herbs can lead to such complications as uterine hypercontractility, uterine rupture, maternal hypotension, and water intoxication (Kelsey & Prevost, 1994).

Several of the herbs in this category are common culinary spices or tea plants that are included due to activity associated with therapeutic dosage. These include Roman chamomile (*Chamaemelum nobile*), saffron (*Crocus sativus*), turmeric (*Curcuma* spp.), lemongrass (*Cymbopogon citratus*), asafetida (*Ferula* spp.), and rosemary (*Rosmarinus officinalis*). The concerns and considerations stated here for the therapeutic use of these are generally not relevant to their common use as beverage teas and spices.

## Therapeutic Use

Midwives of ancient times aided mothers in labor by feeding rye grains contaminated with ergot, a uterine stimulant. They had observed that this course resulted in speedier births and recoveries with far less loss of blood. Many cultures throughout the world continue the use of medicinal plants which act as uterine stimulants, usually in the last trimester of pregnancy and under the direction of a midwife, to facilitate or induce labor, reduce hemorrhage, and relieve the pain of childbirth (Lewis & Elvin-Lewis, 1977).

Contemporary herbal medicine utilizes uterine stimulants to tonify and strengthen the female reproductive organs. Through use of uterine stimulants, women can experience improved reproductive health, including decreased pelvic congestion, improved fertility, more stable pregnancies, and optimal recovery from pelvic infections and invasive gynecologic procedures. Certain of the uterine stimulant herbs are used in limited dosage to prevent miscarriage in the earliest stages of pregnancy. Although this practice has been established over many years, qualified supervision is essential.

Emmenagogues are menstrual inducers. They work as uterine tonics, stimulants, and, in some cases, even as irritants. Properly used during the two weeks preceding the onset of menstruation, emmenagogues normalize menstrual flow. In excessive doses, some emmenagogues function as abortifacients.

## Mechanism of Action

Emmenagogues and uterine stimulants exhibit several different mechanisms of action. Some act directly on the uterus, myometrium, and/or the vasculature surrounding the reproductive organs. Others affect the uterus and reproductive system indirectly, acting through peripheral systems such as the endocrine, cardiovascular, gastrointestinal, or nervous systems.

The direct uterine stimulants are known as oxytocics. Extracts of the bark of the silk tree (*Albizia* spp.) have been shown to stimulate uterine contractions in pregnant mice, rats, guinea pigs, sheep, cows, and monkeys. Similar results have been documented in human uterine preparations *in vitro*. These extracts also reportedly induced partial or complete abortion in rats, rabbits, and guinea pigs (Bingel & Farnsworth, 1980). The active agent of the silk tree is albitocin, which has been shown to be responsible for uterine effects both *in vitro* and *in situ*.

Examples of indirect uterine stimulants include safflower (*Carthamus tinctorius*) and motherwort (*Leonurus sibiricus*). A derivative of *Leonurus*, labdane diterpene, was observed to successfully treat menstrual disturbances (Xu, 1992; Nagasawa, 1992).

The mechanics of each of these herbs are related to the H1-receptor and $\alpha$-adrenergic receptor of the uterus (Shi, 1995).

One class of emmenagogue herbs is described as blood-moving, or as "herbs which invigorate the blood" in the terminology of traditional Chinese medicine. This category includes herbs that increase the menstrual blood flow, not by uterine contraction, but by regulating the blood vessels in the uterus (e.g., yarrow [*Achillea millefolium*]), stimulating the general blood circulation (e.g., prickly ash [*Zanthoxylum* spp.]), or changing the flow characteristics of the uterine blood (e.g., safflower [*Carthamus tinctorius*]) (Ellingwood, 1983; Hoffman, 1983).

Another group of emmenagogue herbs consists of the hormonal regulators. These help to normalize the menstrual flow by regulating the secretions of the hypothalamus, pituitary, and ovaries. An example of this type is *Vitex agnus-castus* which has a specific effect on the pituitary gland (Hobbs, 1990).

**Herbs** listed in the *Botanical Safety Handbook* as emmenagogues or uterine stimulants:

*Achillea millefolium*
*Achyranthes bidentata*
*Albizia julibrissin*
*Angelica archangelica*
*Angelica atropurpurea*
*Artemisia abrotanum*
*Artemisia absinthium*
*Artemisia douglasiana*
*Artemisia lactiflora*
*Artemisia vulgaris*
*Asarum canadense*
*Capsella bursa-pastoris*
*Carthamus tinctorius*
*Caulophyllum thalictroides*
*Chamaelirium luteum*
*Chamaemelum nobile*
*Cimicifuga racemosa*
*Commiphora madagascariensis*
*Commiphora molmol*
*Commiphora myrrha*
*Commiphora mukul*
*Coptis chinensis*
*Corydalis yanhusuo*
*Crocus sativus*
*Curcuma aromatica*
*Curcuma domestica*
*Curcuma longa*
*Cymbopogon citratus*
*Ferula assa-foetida*
*Ferula foetida*

*Ferula rubricaulis*
*Forsythia suspensa*
*Gossypium herbaceum*
*Gossypium hirsutum*
*Hydrastis canadensis*
*Hyssopus officinalis*
*Leonurus cardiaca*
*Leonurus heterophyllus*
*Leonurus sibiricus*
*Levisticum officinale*
*Marrubium vulgare*
*Mentha pulegium*
*Monarda clinopodia*
*Monarda didyma*
*Monarda fistulosa*
*Monarda pectinata*
*Monarda punctata*
*Nardostachys jatamansi*
*Nasturtium officinale*
*Nepeta cataria*
*Petroselinum crispum*
*Polygala senega*
*Rosmarinus officinalis*
*Ruta graveolens*
*Tanacetum vulgare*
*Thuja occidentalis*
*Trillium erectum*
*Vetiveria zizanoides*
*Vitex agnus-castus*
*Ziziphus spinosa*

## GI Irritants

### Definition
GI Irritants are substances that produce irritation and inflammation of the gastrointestinal tract. The most common constituent groups producing GI irritation are saponins, alkaloids, oxalic acid, tannins, and irritative oils.

### Adverse Effects
GI irritants should be used sparingly, especially in individuals with histories of digestive system sensitivity. Excessive use can give rise to severe irritation of the entire digestive tract, including symptoms of chronic abdominal cramps, burning, blistering in the mouth and throat, nausea, colic, and severe watery or bloody diarrhea. GI irritants can also stimulate uterine contractions and therefore should be avoided in pregnancy unless otherwise specified by a qualified health professional. Some irritants may aggravate GI ulcers. Those herbs containing oxalic acid, such as quillaja (*Quillaja saponaria*) and skunk cabbage (*Symplocarpus foetidus*), may promote kidney irritation and the progress of stones See the constituent section on oxalates in these appendices for further information.

### Therapeutic Use
Although GI irritants can be deliberately used to stimulate functioning of the gastrointestinal tract, most often irritation is an undesirable side effect associated with the herbs listed in this category and many other substances.

**Herbs** listed in the *Botanical Safety Handbook* as GI irritants:

*Arctostaphylos uva-ursi*
*Cetraria islandica*
*Chamaelirium luteum*
*Coffea arabica*
*Cola acuminata*
*Cola nitida*
*Quillaja saponaria*
*Ruta graveolens*
*Sanguinaria canadensis*
*Schinus terebinthifolia*
*Schinus molle*
*Symplocarpus foetidus*
*Trillium erectum*

APPENDIX 2: HERBAL ACTION PROFILES

## *Monoamine Oxidase (MAO) Interaction*

## Definition

Monoamine oxidase (MAO) enzymes are responsible for the breakdown of the neurotransmitters dopamine, seratonin, epinephrine, and norepinehrine. Pharmaceutical monoamine oxidase inhibitors (MAOIs), prescribed for depression and anxiety, temporarily disrupt these enzymes; thereby creating an increase in neurotransmitters.

## Adverse Effects

There is some concern regarding potentially harmful interactions from simultaneous consumption of pharmaceutical MAOIs and any of the herbs listed here.

Standard texts report severe hypertensive reactions when MAOIs are prescribed simultaneously with sympathomimetic agents, including ephedrine, an alkaloid found in *Ephedra*. One of the common side effects observed for prescription MAOIs is orthostatic hypotension, a condition characterized by lowered blood pressure when changing from a supine to an erect position. This is also reported as a primary effect of yohimbine hydrochhloride, leading to a concern that the concurrent use of yohimbe (*Pausinystalia yohimbe*) might aggravate such a response (Reynolds, 1989).

The cautions associated with consumption of nutmeg or mace (*Myristica fragrans*) and St. John's wort (*Hypericum perforatum*) while using MAOIs are somewhat speculative. It has been proposed that myristicin, an alkenylbenzene present in *Myristica*, converts into an amine neurotransmitter-like form, and so may potentiate MAO inhibition. A similar potential for augmenting the physiological activity of prescription MAOIs is attributed to *Hypericum*, since the well-known antidepressant activity of this herb is believed to be due to an MAO-inhibiting action similar to the pharmaceutical substances (Bruneton, 1995; Leung & Foster, 1996).

## Therapeutic Use

A primary use of St. John's wort has been for treatment of symptoms associated with depression and other emotional states for at least 275 years, as evidenced by its classification as an agent for madness and melancholy as early as 1719 (Anonymous, 1719). Contemporary internal usage has been established for depression, anxiety and nervousness and antiretroviral activity has been observed for hypericin. External application of an oil extract is recorded for various injuries to the skin (Wichtl, 1994) See the *Photosensitising* section of this Appendix (*page 176*) for further information on the use of this herb.

Therapeutic use of *Ephedra* is discussed in the *Nervous System Stimulant* section of this Appendix (*page 175*). The therapeutic uses of the other herbs listed here are not so closely related to concerns for MOA interaction and so have been excluded.

## Mechanism of Action

The mechanisms of action of each of these herbs, insofar as they are understood and as they are related to monoamine oxidase and MAOIs, have been presented above in the discussion of *Adverse Effects*.

**Herbs** listed in the *Botanical Safety Handbook* with potential monoamine oxidase interaction:

*Ephedra distachya*  
*Ephedra equisetina*  
*Ephedra gerardiana*  
*Ephedra intermedia*  
*Ephedra sinica*  

*Eschscholzia californica*  
*Ginkgo biloba*  
*Hypericum perforatum*  
*Myristica fragrans*  
*Pausinystalia yohimbe*

# Nervous System Stimulants

## Definition

Nervous system stimulants are substances which increase the activity of some portion of the brain or of the spinal cord. Although there are several pharmacological classifications of stimulants, only the xanthine derivatives and ephedrine are relevant to the botanical stimulants listed here.

## Adverse Effects

Nervous system stimulants are not recommended for excessive or long-term use. Side effects include nervousness, anxiety, heart irregularities, headaches, tremors, hypertension, restlessness, insomnia, daytime irritability, irritation of the gastric mucosa, and diuresis (Feldman & Davidson, 1986).

There are also less immediately noticeable effects of each of the plants listed in this section. All of the stimulant beverages may irritate the digestive mucosa in some persons, since they contain high concentrations of tannins or irritating essential oils. There has been some concern stated regarding the effect of long-term use of stimulants on the function of the adrenal gland (Friedman, 1979).

Because its use is so widespread, the study of the effects of caffeine has been actively pursued. Conclusive results, however, are controversial in a wide range of studies on caffeine's role in cholesterol levels, fibrocystic breast conditions, heart disease, and cancer. The U.S. FDA since 1980 has cautioned pregnant women to eliminate or moderate caffeine intake, and though no regulatory or label restrictions have been mandated, moderation is prudent, especially for those with a history of difficulty in conception.

Concerns have surfaced in the last several years regarding the use of some of these herbs, especially *Ephedra* and its alkaloids, for non-traditional uses.

The FDA has compiled data which purportedly show an associative relationship between consumption of insignificant quantities of ephedrine alkaloids and serious side effects, including death (Anonymous, 1996). This compilation, however, has not been reviewed by unbiased and qualified experts. Nonetheless, there is a real concern for products that contain these stimulant alkaloids and are pointedly marketed as substitutes for illicit drugs. In the light of this potential and actual misuse, AHPA has established a position recommending cautionary labeling and dosage limitations for products containing *Ephedra*, the details of which appear in the body of this text (*see page 46*). Many states have placed limitations on the sale of ephedrine containing products. It is likely that the FDA will announce new regulations for the sale of *Ephedra* sometime in 1997.

## Therapeutic Use

Nervous system stimulants have long been used to provide a sense of exhilaration and mental clarity and to alleviate fatigue and drowsiness. Most ancient cultures used stimulant herbs both ritualistically and in everyday life. For instance, native tribes of South America have long chewed coca leaves, the botanical source of cocaine. This use provides a long-lasting, low-grade euphoria that reduces appetite, increases physical stamina, and counters symptoms associated with oxygen deprivation. Nervous system stimulants from other cultures include coffee, kola nut, ephedra, yerba maté, yohimbe, and tea (Gilbert, 1976).

In contemporary western culture, the most popular stimulant is the alkaloid caffeine, which is readily available in coffee, tea, and many carbonated beverages, and in over-the-counter medications. The consumption of stimulant beverages is nearly universal, with daily ingestion of 200 mg of caffeine, the equivalent of three cups of coffee, reported for over one third of Americans (Narod et al., 1991).

Additional therapeutic values are associated with all of the stimulants included in this review. Various species of *Ephedra*, as well as the contained alkaloid ephedrine and its synthetic derivative, are used for the symptomatic relief of congestive bronchial conditions. All of the caffeine-containing herbs exhibit diuretic activity. Caffeine in its pure form is used in the treatment of apnea in infants and, in combination with other drugs, has analgesic uses.

## Mechanism of Action

The stimulant properties of herbal nervous system stimulants are derived from their contained alkaloids. In general, these stimulants act on the central nervous system and the adrenal glands, increasing the synthesis and release of specific neurotransmitters and hormones. Caffeine, for example, increases the secretion of the neurotransmitter norepinephrine and the adrenal hormone epinephrine (adrenaline), while blocking central adenosine receptors. Ephedrine acts as a sympathomimetic agent, with direct and indirect effects on both α and β adrenergic receptors. A number of physiological responses are effected, including peripheral vasoconstriction and cardiac stimulation, bronchodilation and stimulation of the respiratory center, and others (Gawin, 1988; Gilman et al., 1985).

**Herbs** listed in the Botanical Safety Handbook as nervous system stimulants:

*Camellia sinensis*
*Coffea arabica*
*Cola acuminata*
*Cola nitida*
*Ephedra distachya*
*Ephedra equisetina*
*Ephedra gerardiana*
*Ephedra intermedia*
*Ephedra sinica*
*Ilex paraguayensis*
*Paullinia cupana*

## *Photosensitizing*

### Definition
Contributing to a condition caused by the ingestion of specific substances in association with exposure to sunlight that results in various skin irritations, such as rashes, swelling, and hyperpigmentation.

### Adverse effects
Phototoxic reactions are exhibited by skin cell damage characterized by rashes and blisters and by bizarre hyperpigmentation. Phototoxic reactions are generally associated with products containing furocoumarins, or psoralens. These phenolic compounds are found in more than two dozen plant sources, including species of the Rutaceae, Apiaceae, Fabaceae, and Moraceae families, and in all of the plants included in this listing. Development of photosensitivity is dependent upon the concentration of the compounds contained in the plant. However, photosensitivity associated with the use of some common herbs such as *Angelica* and *Hypericum* only occurs at doses many times higher than standard recommended dosages. Similarly, fair-skinned individuals are more susceptible to developing photosensitivity than others.

Photoallergic reactions are less common than phototoxic reactions and occur in genetically predisposed individuals. Often the amount of plant substance required to elicit a response is quite small. Photoallergic reactions are exhibited as skin conditions characteristic of allergic contact dermatitis.

# APPENDIX 2: HERBAL ACTION PROFILES

## Therapeutic use

The therapeutic uses and value of the various substances included here that exhibit a photosensitizing side effect are specific to the individual herbs, and there is no particular therapeutic intent related to these herbs as a group. Both St. John's wort (*Hypericum*) and psoralea (*Cullen*) are used as treatments for various skin diseases. A therapy for the treatment of severe psoriasis and other skin conditions, known as PUVA of photochemotherapy, relies on the photosensitizer methoxsalen, a psoralen derived from the fruit of *Ammi majus*, for its ability to increase skin reactivity to long-wavelength ultraviolet radiation (Reynolds, 1989).

## Mechanism of Action

There are two distinct types of photosensitizing reactions associated with the use of photosensitizing agents. Phototoxic reactions occur by light activation of plant substances such as furocoumarins. This activation results in free radical formation that leads by means of photochemical reactions to phototoxic effects, usually manifest as an extreme sunburn or second degree burns. Photoallergic reactions, most often caused by topical agents, occur by light activation of a photosensitizing molecule in a light sensitive allergen. This causes the formation of a photohapten. When the photohapten conjugates with a suitable protein in the skin, a photoantigen is produced; leading to photoallergic effects such as urticaria, rash, eczema, and sunburn (Katcher et al., 1983).

**Herbs** listed in the *Botanical Safety Handbook* with potential photosensitizing action:

*Angelica archangelica*  
*Angelica atropurpurea*  
*Angelica pubescens*  
*Apium graveolens*  

*Citrus aurantium*  
*Cullen corylifolia*  
*Hypericum perforatum*  
*Ruta graveolens*

---

## *Stimulant Laxatives*

### Definition

Stimulant laxatives are agents used to relieve constipation by local stimulation and contraction of the smooth muscle of the lower bowel. This results in increased peristalsis that empties stools more quickly.

### Adverse effects

Short-term side effects of stimulant laxative consumption may include intestinal cramps, uterine contractions, and watery diarrhea. Long-term use can cause dependency, resulting in a significant reduction in the ability to perform normal bowel functions without the aid of laxatives. When used in excess or for long periods, the resultant loss of fluids and electrolytes, especially potassium, can cause pathological alterations to the colon, renal malfunction, or heart palpitations. Patients taking cardiac glycosides are particularly susceptible to this last stated concern (De Smet, 1993).

Labeling required by the FDA's current tentative final monograph for over-the-counter stimulant laxatives includes those containing aloe, cascara sagrada (*Rhamnus purshiana*), castor oil (from *Ricinus communis*), or derivatives of senna (*Cassia* spp.). Such products must include the following warnings:

DO NOT USE LAXATIVE PRODUCTS WHEN ABDOMINAL PAIN, NAUSEA, OR VOMITING ARE PRESENT UNLESS DIRECTED BY A DOCTOR.

IF YOU HAVE NOTICED A SUDDEN CHANGE IN BOWEL HABITS THAT PERSISTS OVER A PERIOD OF 2 WEEKS, CONSULT A DOCTOR BEFORE USING A LAXATIVE.

LAXATIVE PRODUCTS SHOULD NOT BE USED FOR A PERIOD LONGER THAN 1 WEEK UNLESS DIRECTED BY A DOCTOR.

RECTAL BLEEDING OR FAILURE TO HAVE A BOWEL MOVEMENT AFTER USE OF A LAXATIVE MAY INDICATE A SERIOUS CONDITION. DISCONTINUE USE AND CONSULT YOUR DOCTOR.

An emergency regulation issued by the California Department of Health Services effective on January 1, 1997 requires a label statement for supplements containing the dried latex of aloe, buckthorn berries (*Rhamnus catharticus*), cascara sagrada bark (*Rhamnus purshiana*), frangula bark (*Rhamnus frangula*), rhubarb root (*Rheum* spp.), and/or senna (*Cassia* spp.):

NOTICE: THIS PRODUCT CONTAINS (NAME OF SUBSTANCE(S) AND COMMON NAME IF DIFFERENT). READ AND FOLLOW DIRECTIONS CAREFULLY. **DO NOT USE IF YOU HAVE OR DEVELOP DIARRHEA, LOOSE STOOLS, OR ABDOMINAL PAIN.** CONSULT YOUR PHYSICIAN IF YOU HAVE FREQUENT DIARRHEA, ARE PREGNANT, NURSING, TAKING MEDICATION, OR HAVE A MEDICAL CONDITION.

The regulation requires specific formatting and specifically exempts aloe vera juice, gel and concentrations thereof.

## Therapeutic use

Stimulant laxative plants produce a laxative or cathartic effect. Depending on the dose and plant used, a soft or fluid stool is produced.

## Mechanism of action

The action of stimulant laxative herbs is, in most cases, due primarily to their content of anthraquinones. The one exception among the herbs listed in this category is castor oil (from *Ricinus communis*), the action of which is due to ricinoleic acid.

Stimulant laxatives increase the motility of the colon, induce changes in the surface cells of the colon, and cause the loss of water and electrolytes. Although intensive research has been performed, the mechanism of action is still unclear. However, herbs that contain anthraquinones do affect the colonic mucosa and produce a laxative effect. Anthraquinones disturb the equilibrium between the absorption of water from the intestinal lumen via an active sodium transport and the secretion of water into the lumen by the hydrostatic blood pressure or a prostaglandin-dependent chloride secretion. Anthraquinone-glycosides utilize the intestinal flora to produce a laxative effect and are stronger in action than anthraquinone-glycones, which are absorbed in the stomach and duodenum (De Smet, 1993). Both aglycones and glycosides occur in *Aloe, Cassia, Rhamnus,* and *Rheum*.

**Herbs** listed in the *Botanical Safety Handbook* as stimulant laxatives:

*Aloe ferox*
*Aloe perryi*
*Aloe vera*
*Rhamnus catharticus*
*Rhamnus frangula*
*Rhamnus purshiana*
*Rheum officinale*

*Rheum palmatum*
*Rheum tanguticum*
*Ricinus communis*
*Senna tora*
*Senna alexandrina*
*Senna obtusifolia*

# Appendix 3

## HERB LISTINGS BY CLASSIFICATION

Note: Please check the text for part of herb classified.

*The following herbs are listed in the* **Botanical Safety Handbook** *in* **Class 2a**:
For external use only, unless otherwise directed by an expert qualified in the appropriate use of this substance.

## Herbs by Common Name

alkanet
borage
male fern
Joe Pye

henna
Russian comfrey
comfrey

## Herbs by Latin Name

*Alkanna tinctoria*
*Borago officinalis*
*Dryopteris filix mas*
*Eupatorium purpureum*
*Lawsonia inermis*

*Symphytum asperum*
*Symphytum x uplandicum*
*Symphytum officinale*

# APPENDIX 3: HERB LISTINGS BY CLASSIFICATION

*The following herbs are listed in the* **Botanical Safety Handbook** *in* **Class 2b**: Not to be used during pregnancy unless otherwise directed by an expert qualified in the appropriate use of this substance.

## Herbs by Common Name

achyranthes
alkanet
aloe vera
American liverleaf
American pennyroyal
andrographis
angelica
anise
arnica
asafetida
asarabacca
ashwagandha
barberry
barley
basil
beebalm
beth root
black cohosh
bladderwrack
blessed thistle
bloodroot
blue cohosh
blue flag
blue lobelia
blue vervain
borage
buchu
bugleweed
California poppy
California spikenard
camphor
Canada snakeroot
capillary artemisia
cascara sagrada
cassia
castor
catnip
celandine

celery
chaste tree
chervil
Chinese goldthread
Chinese motherwort
Chinese rhubarb
coffee
coltsfoot
comfrey
cornflower
corydalis
cotton
culver's root
cyathula
dong quai
dyer's broom
elecampane
ephedra
European pennyroyal
European vervain
fenugreek
feverfew
forsythia
fritillary
garlic
ginger
goldenseal
goldthread
guggul
horehound
hyssop
inmortal
ipecac
Japanese arisaema
jatamansi
Job's tears
Joe Pye
jujube seeds

## Class 2b:
## Herbs by Common Name
*Continued*

juniper
lemongrass
licorice
lobelia
lomatium
lovage
lycium
mace
magnolia
maidenhair fern
male fern
ming dang shen
motherwort
mugwort
myrrh
nutmeg
ocotillo
Oregon grape
osha
parsley
peach
phellodendron bark
pinellia
pleurisy root
prickly ash
psoralea
purging buckthorn
purslane
quassia
quinine
red clover

Roman chamomile
rosemary
rue
safflower
saffron
sage
san qi ginseng
scouring rush
seneca snakeroot
senna
sete sangrias
shepherd's purse
Sichuan lovage
Sichuan pepper
silk tree
small spikenard
southernwood
spikenard
sweet annie
thuja
tree peony bark
trichosanthes
turmeric
uva-ursi
vetiver
watercress
wild carrot
wild indigo
wormwood
yarrow
zedoary

APPENDIX 3: HERB LISTINGS BY CLASSIFICATION

*The following herbs are listed in the* **Botanical Safety Handbook** *in* **Class 2b**: Not to be used during pregnancy unless otherwise directed by an expert qualified in the appropriate use of this substance.

## Herbs by Latin Name

*Achillea millefolium*
*Achyranthes bidentata*
*Acorus calamus*
*Acorus gramineus*
*Adiantum pedatum*
*Albizia julibrissin*
*Alkanna tinctoria*
*Aloe ferox*
*Aloe perryi*
*Aloe vera*
*Andrographis paniculata*
*Angelica archangelica*
*Angelica atropurpurea*
*Angelica sinensis*
*Anthriscus cerefolium*
*Apium graveolens*
*Aralia californica*
*Aralia nudicaulis*
*Aralia racemosa*
*Arctostaphylos uva ursi*
*Arisaema japonicum*
*Aristolochia clematitis*
*Aristolochia contorta*
*Aristolochia debilis*
*Aristolochia serpentaria*
*Arnica latifolia*
*Arnica montana*
*Arnica* spp.
*Artemisia abrotanum*
*Artemisia absinthium*
*Artemisia annua*
*Artemisia capillaris*
*Artemisia douglasiana*
*Artemisia lactiflora*
*Artemisia scoparia*
*Artemisia vulgaris*
*Asarum canadense*
*Asarum europaeum*
*Asarum heteropides*

*Asarum sieboldii*
*Asclepias asperula*
*Asclepias tuberosa*
*Baptisia tinctoria*
*Barosma betulina*
*Barosma crenulata*
*Barosma serratifolia*
*Berberis vulgaris*
*Borago officinalis*
*Capsella bursa-pastoris*
*Carthamus tinctorius*
*Caulophyllum thalictroides*
*Cephaelis ipecacuanha*
*Chamaelirium luteum*
*Chamaemelum nobile*
*Changium smyrnoides*
*Chelidonium majus*
*Cimicifuga racemosa*
*Cinchona calisaya*
*Cinchona ledgeriana*
*Cinchona officinalis*
*Cinchona pubescens*
*Cinnamomum camphora*
*Cinnamomum cassia*
*Cinnamomum verum*
*Cnicus benedictus*
*Coffea arabica*
*Coix lacryma-jobi*
*Cola acuminata*
*Cola nitida*
*Commiphora madagascariensis*
*Commiphora molmol*
*Commiphora mukul*
*Commiphora myrrha*
*Coptis chinensis*
*Coptis groenlandica*
*Corydalis yanhusuo*
*Crocus sativus*
*Cullen corylifolia*

Class 2b:
## Herbs by Latin Name
Continued

Cuphea balsamona
Curcuma aromatica
Curcuma domestica
Curcuma longa
Curcuma zedoaria
Cyathula officinalis
Cymbopogon citratus
Daucus carota
Dryopteris filix mas
Ephedra distachya
Ephedra equisetina
Ephedra gerardiana
Ephedra intermedia
Ephedra sinica
Equisetum hyemale
Eschscholzia californica
Eupatorium purpureum
Ferula assa-foetida
Ferula foetida
Ferula rubricaulis
Forsythia suspensa
Fouquieria splendens
Fritillaria cirrhosa
Fritillaria thunbergii
Fucus vesiculosus
Genista tinctoria
Glycyrrhiza echinata
Glycyrrhiza glabra
Glycyrrhiza uralensis
Gossypium herbaceum
Gossypium hirsutum
Hedeoma pulegioides
Hepatica nobilis
Hordeum vulgare
Hydrastis canadensis
Hyssopus officinalis
Inula helenium
Iris versicolor
Iris virginica
Juniperus communis
Juniperus monosperma

Juniperus osteosperma
Juniperus oxycedrus
Juniperus virginiana
Leonurus cardiaca
Leonurus heterophyllus
Leonurus sibiricus
Leptandra virginica
Levisticum officinale
Ligusticum chuanxiong
Ligusticum porteri
Lobelia inflata
Lobelia siphilitica
Lomatium dissectum
Lycium barbarum
Lycium chinense
Lycopus americanus
Lycopus europaeus
Lycopus virginicus
Magnolia officinalis
Mahonia aquifolium
Mahonia nervosa
Mahonia repens
Marrubium vulgare
Mentha pulegium
Monarda clinopodia
Monarda didyma
Monarda fistulosa
Monarda pectinata
Monarda punctata
Myristica fragrans
Nardostachys jatamansi
Nasturtium officinale
Nepeta cataria
Paeonia suffruticosa
Panax notoginseng
Petroselinum crispum
Phellodendron amurense
Phellodendron chinense
Picrasma excelsa
Pilocarpus jaborandi
Pilocarpus microphyllus

## Class 2b:
## Herbs by Latin Name
*Continued*

*Pilocarpus pennatifolius*
*Pimpinella anisum*
*Pinellia ternata*
*Podophyllum hexandrum*
*Podophyllum peltatum*
*Polygala senega*
*Portulaca oleracea*
*Prunus persica*
*Quassia amara*
*Rhamnus catharticus*
*Rhamnus frangula*
*Rhamnus purshiana*
*Rheum officinale*
*Rheum palmatum*
*Rheum tanguticum*
*Ricinus communis*
*Rosmarinus officinalis*
*Ruta graveolens*
*Salvia officinalis*
*Sanguinaria canadensis*
*Senna alexandrina*
*Senna obtusifolia*

*Senna tora*
*Symphytum officinale*
*Tanacetum parthenium*
*Tanacetum vulgare*
*Thuja occidentalis*
*Trichosanthes kirilowii*
*Trifolium pratense*
*Trigonella foenum-graecum*
*Trillium erectum*
*Tussilago farfara*
*Verbena hastata*
*Verbena officinalis*
*Vetiveria zizanoides*
*Vitex agnus-castus*
*Withania somnifera*
*Zanthoxylum americanum*
*Zanthoxylum bungeanum*
*Zanthoxylum clava-herculis*
*Zanthoxylum schinifolium*
*Zanthoxylum simulans*
*Zingiber officinale*
*Ziziphus spinosa*

*The following herbs are listed in the* **Botanical Safety Handbook** *in* **Class 2c**: Not to be used while nursing unless otherwise directed by an expert qualified in the appropriate use of this substance.

## Herbs by Common Name

alkanet
aloe vera
aloes
basil
black cohosh
bladderwrack
borage
bugleweed
cascara sagrada
Chinese rhubarb
coltsfoot

comfrey
elecampane
ephedra
garlic
Joe Pye
licorice
male fern
purging buckthorn
senna
stillingia
wormwood

## Herbs by Latin Name

*Alkanna tinctoria*
*Allium sativum*
*Aloe ferox*
*Aloe perryi*
*Aloe vera*
*Artemisia absinthium*
*Borago officinalis*
*Cimicifuga racemosa*
*Dryopteris filix mas*
*Ephedra distachya*
*Ephedra equisetina*
*Ephedra gerardiana*
*Ephedra intermedia*
*Ephedra sinica*
*Eupatorium purpureum*
*Fucus vesiculosus*
*Glycyrrhiza echinata*

*Glycyrrhiza glabra*
*Glycyrrhiza uralensis*
*Inula helenium*
*Lycopus americanus*
*Lycopus europaeus*
*Lycopus virginicus*
*Rhamnus catharticus*
*Rhamnus frangula*
*Rhamnus purshiana*
*Rheum officinale*
*Rheum palmatum*
*Rheum tanguticum*
*Senna obtusifolia*
*Senna tora*
*Stillingia sylvatica*
*Symphytum officinale*
*Tussilago farfara*

# APPENDIX 3: HERB LISTINGS BY CLASSIFICATION

*The following herbs are listed in the* **Botanical Safety Handbook** *in* **Class 3**:
Herbs for which significant data exist to recommend the following labeling: "To be used only under the supervision of an expert qualified in the appropriate use of this substance." Labeling must include proper use information: dosage, contraindications, potential adverse effects and drug interactions, and any other relevant information related to the safe use of the substance.

## Herbs by Common Name

aconite
American hellebore
American mistletoe
apricot
belladonna
bitter almond
boxwood
calamus
Chinatree
digitalis
eastern red cedar
germander
golden eye grass
grass leaved calamus
Himalayan mayapple
jaborandi
jalap
lily of the valley

long birthwort
ma dou ling
Madagascar periwinkle
mandrake
mayapple
peach
poke
pomegranate
qing mu xiang
scopolia
Scotch broom
Sichuan pagoda tree
spreading dogbane
tansy
tian xian teng
tonka
Virginia snakeroot
wahoo

## Herbs by Latin Name

*Aconitum carmichaelii*
*Acorus calamus*
*Acorus gramineus*
*Apocynum androsaemifolium*
*Apocynum cannabinum*
*Aristolochia clematitis*
*Aristolochia contorta*
*Aristolochia debilis*
*Aristolochia serpentaria*
*Arnica latifolia*
*Arnica montana*
*Arnica spp.*
*Atropa belladonna*
*Buxus sempervirens*

*Catharanthus roseus*
*Convallaria majalis*
*Curculigo orchioides*
*Cytisus scoparius*
*Digitalis purpurea*
*Dipteryx odorata*
*Dipteryx oppositifolia*
*Dryopteris filix-mas*
*Euonymus atropurpureus*
*Ipomoea purga*
*Juniperus virginiana*
*Mandragora officinarum*
*Melia azedarach*
*Melia toosendan*

Class 3:
## Herbs by Latin Name
Continued

Myristica fragrans
Phoradendron leucarpum
Phytolacca americana
Pilocarpus jaborandi
Pilocarpus microphyllus
Pilocarpus pennatifolius
Podophyllum hexandrum
Podophyllum peltatum

Prunus armeniaca
Prunus dulcis
Prunus persica
Punica granatum
Scopolia carniolica
Tanacetum vulgare
Teucrium chamaedrys
Veratrum viride

# Primary References

All of the toxicity data in each of the following references are included in the *Botanical Safety Handbook*. In choosing these references the Editors sought to include the most authoritative contemporary texts as well as historical data. These primary references encompass information regarding herb usage, toxicology, and regulatory status throughout Europe and in China, Australia, and North America.

1. Baker, C., Chair. 1990. *Report of the South Australian Working Party on Natural and Nutritional Substances*. South Australian Working Party on Natural and Nutritional Substances, South Australian Health Commission.

    This document represents the recommendations made by the expert advisory panel chartered in 1985 to address issues related to the implementation of the Controlled Substances Act of 1984. The Working Party developed a three-tiered classification system based on toxicity and known risk factors, and a label warning mechanism when appropriate. Any recommended usage limitation or cautionary labeling is reported in the *Botanical Safety Handbook*.

2. Bensky, D. and A. Gamble. 1986. *Chinese Herbal Medicine*. Seattle: Eastland Press, Inc.

    This work (as well as its second edition published in 1993), translated from contemporary Chinese texts, is the most authoritative compilation of information available in English about commonly used Chinese therapeutic herbs. In describing both traditional applications and pharmacological activity the authors successfully bridge the gap between Eastern and Western therapeutic theories. Each listing contains a section on *Cautions & Contraindications* and, when necessary, a *Toxicity* note. This reference provides the primary reference for **Standard Dose** when required for Chinese herbs.

3. Blackburn, J.L., Chair. 1993. *Second Report of the Expert Advisory Committee on Herbs and Botanical Preparations to the Health Protection Branch, Health Canada*. Ministry of Health Canada.

    The Expert Advisory Committee was established in 1984 to assess safety concerns associated with the sale of herbs and botanicals as foods. Its work resulted in a recommendation that a class of Traditional Medicines be identified, an action that has subsequently been adopted in Canada. This work classifies 64 plants as adulterants in foods and 7 others as requiring labeling which contraindicates in pregnancy. This list was later modified (Welsh, 1995), and the current classifications are reported in the *Botanical Safety Handbook*.

4. Blumenthal, M., ed.; S. Klein, trans. 1993. *Therapeutic Monographs on Medicinal Plants for Human Use* of the Commission E Special Expert Committee, Federal Health Agency, Germany (Draft). Austin: American Botanical Council.

   The Editors had the good fortune to have a Draft translation of these monographs made available by the American Botanical Council. The monographs were developed by an expert committee formed by the Federal Health Agency of the Federal Republic of Germany to assess both safety and efficacy of herbs, which are classified in Germany as drugs. Information is included on constituents, contraindications, side effects, risks and drug interactions, use, action, dosage and mode of administration; and finally, when the committee determined that a benefit had not been documented or was insufficient in the light of a known risk, an evaluation to that effect is included. Though this compilation proved to be a valuable, if overcautious resource, it unfortunately does not list any references, thus preventing a review of the accuracy of its data.

5. Bradley, P.R., ed. 1992. *British Herbal Compendium, volume 1*. Dorset: British Herbal Medicine Association.

   The British Herbal Medicine Association continues to provide high quality and informative works addressed to a thorough knowledge of herbal therapeutics. Extremely well referenced, the *Compendium* includes pharmacological data as well as clinical information such as actions, indications, contraindications, dosage, and toxicology. Unfortunately there are only 84 entries in this volume and many of the most widely used herbs in North America are not included.

6. Chadha, Y.R. et al., eds. 1952-88. *The Wealth of India* (Raw Materials), 11 vols. New Delhi: Publications and Information Directorate, CSIR.

   Though some of the volumes are dated, *Wealth of India* is an invaluable source of information for a very large number of medicinal plants. Each plant monograph includes botanical, constituent, and pharmacological information. Toxicological data and adverse effects are included for any herb which has been reported to present such concerns.

7. Chang, H.M. and P.P.H. But. 1986. *Pharmacology and Applications of Chinese Materia Medica*. Philadelphia: World Scientific.

   This two-volume set (now out of print) was compiled by a team of pharmacologists and clinicians who combed the available literature to develop a comprehensive pharmacology text for Chinese herbal medicine. Covering just under 200 herbs, each monograph contains notes on historical use, chemical composition, pharmacodynamics, clinical reports, and toxicity. Much of the toxicological data, however, was extrapolated from studies involving laboratory animals injected with herb extracts or isolated constituents.

8. De Smet, P.A.G.M. et al., eds. 1992. *Adverse Effects of Herbal Drugs 1*. New York: Springer-Verlag.

    Well-written, researched and referenced, and seemingly well-intentioned, the *Adverse Effects* series provided valuable information to the Editors. This volume discusses 22 plants in depth, with an emphasis on profiling adverse reactions and examining appropriate use during pregnancy and lactation. Westendorf's thorough analysis of the concerns associated with pyrrolizidine alkaloids is a significant reference for these constituents.

9. De Smet, P.A.G.M. et al., eds. 1993. *Adverse Effects of Herbal Drugs 2*. New York: Springer-Verlag.

    Volume 2 of *Adverse Effects of Herbal Drugs* provides the basis for a complete understanding of the anthraquinone-containing stimulant laxative herbs. Besides the addition of another 20 plants to the series, an introductory chapter on international regulatory policies provided supplementary data for some of the herbs in the *Botanical Safety Handbook*.

10. Felter, H.W. and J.U. Lloyd. 1898. *King's American Dispensatory*. Cincinnati: The Ohio Valley Co.

    One of the most informative surviving texts of the Eclectic physicians of the late 19th and early 20th century, this reference is invaluable in preserving serious analysis of hundreds of botanicals. The authors were writing to inform the practicing physician, so data are included on dosage and indications, and side effects are recorded based on the observations of the authors as well as on correspondence with practitioners, thus preserving the records of extensive clinical experience with these herbs.

11. Frone, D. and H. Pfänder. 1983. *A Colour Atlas of Poisonous Plants*. London: Wolfe Publishing.

    A high-quality modern work on poisonous plants. Well-formatted for ease of use and includes color photographs of plant parts, making it useful for identification. Descriptions are accurate and well-organized. Treatment options are called out in separate boxes, and valuable microscopic characters are included. A detailed reference list emphasizes the authors' thorough review of the European and world literature on poisonous plants. Case histories of human poisoning incidents are also valuable.

12. Hardin, J.W. and J.M. Arena. 1974. *Human Poisoning from Native and Cultivated Plants*, 2nd ed. Durham, NC: Duke University Press.

    This text is primarily concerned with toxicological concerns associated with inadvertent consumption of poisonous plants, especially by children. Instructions to physicians for emergency treatment are included for many of the listed plants. Though much of the focus of this work is not relevant to the consumption of dietary supplements, and thus to the present work, this manual provides basic and valuable information for the truly poisonous plants which are encountered in North America.

13. Hsu, H. 1986. *Oriental Materia Medica, A Concise Guide*. Long Beach: Oriental Healing Arts.

> The distinguished principal author of this document has been an active participant in the field of Chinese herbal medicine since 1945. The usage of each herb is described within the conceptual framework of Traditional Chinese Medicine, and each listing includes data related to chemical constituents (though not quantified), pharmacology, and dosage. This reference provides useful support to Bensky & Gamble's *Chinese Herbal Medicine* in examining Chinese herbs.

14. Huang, K.C. 1993. *The Pharmacology of Chinese Herbs*. Boca Raton, FL: CRC Press.

> This manual, written by a pharmacologist trained primarily in Western traditions, emphasizes the chemistry of Chinese medicinal plants while presenting their actions in Western medical terms. Its value as a reference for the present work is in its information on chemical constituents and in its discussions on toxicity for those herbs which present toxic concerns.

15. Kingsbury, J.M. 1964. *Poisonous Plants of United States and Canada*. Englewood Cliffs, NJ: Prentice-Hall, Inc.

> Kingsbury's work is a classic text, often based on reports of livestock poisoning. The book covers the chemistry and pharmacology of numerous plants and thoroughly reviews the literature before the 1960s.

16. Lampe, K.F. and M.A. McCann. 1985. *AMA Handbook of Poisonous and Injurious Plants*. Chicago: AMA.

> A concise manual of common poisonous plants likely to be encountered in the home and field. The abbreviated sections on each plant and the extensive color plate section makes the work more of a field guide to poisonous plants. The work includes extensive alternate common and Latin names for each plant as well as detailed descriptions, aiding in plant identification. Supportive treatment measures are included.

17. Leung, A.Y. and S. Foster. 1996. *Encyclopedia of Common Natural Ingredients Used in Food, Drugs and Cosmetics*, 2nd ed. New York: John Wiley & Sons, Inc.

> The revised edition of this essential work was published during the first editing phase of the *Botanical Safety Handbook*. The revisions from the original, published in 1980 and included as a primary reference when the present work was initiated, are significant, both in broadening the scope to include numerous additional plants and in updating information on the original listings. This timely publication provides an exceedingly well-referenced and well-organized review of the plants most commonly encountered in the North American marketplace. Quantified constituent data as well as useful information on toxicity are included, the latter discussed both from a traditional and from a contemporary chemical perspective.

18. List, P.H., and L. Hörhammer. 1973-'79. *Hagers Handbuch der Pharmazeutischen Praxis*, 7 vols. New York: Springer-Verlag.

    *Hagers Handbuch* contains some of the most accurate and comprehensive information on herbal medicine in any language. In fact, many of our other primary references cite *Hagers* as one of their information sources. The listing for each plant is exhaustive in addressing the botany, macroscopic and microscopic descriptions, constituent profile, pharmacological data, toxicology, clinical studies, dosage, etc. Translation of this essential reference was provided by Christopher Hobbs, L.Ac.

19. Muenscher, W.C. 1951. *Poisonous Plants of the United States*. New York: Macmillan Company.

    Primarily a manual for those plants which present some danger to grazing livestock, this mid-century reference nonetheless provides a useful adjunct to Hardin & Arena's *Human Poisoning from Native and Cultivated Plants* in the examination of the highly toxic plants in nature.

20. Office of the Federal Register, publ. 1994. *Code of Federal Regulations 21: Food and Drugs*. Washington: U.S. Government Printing Office.

    This annual publication is the official reference for governmental regulations as they relate to plant source food and drug ingredients. While there are occasional labeling requirements and use restrictions proscribed for certain pharmacologically active herbs, the primary information transcribed to the *Botanical Safety Handbook* is the usage limitations listed in §172.510 for natural flavoring agents of botanical origin.

21. Reynolds, J.E., ed. 1989. *Martindale: The Extra Pharmacopoeia*, 29th ed. London: The Pharmaceutical Press.

    In continuous publication since 1883, *Martindale* is an extensive cross-reference of over 5000 drugs, proprietary medicines, and related substances written as an instructive resource for physicians and pharmacists. Although herbs do not comprise a significant portion of this work, the inclusion of data from 28 contemporary pharmacopoeias and the editors' attentions to adverse effects make this a useful secondary reference. Thorough examinations of the absorption and fate of various classes of pharmaceutical agents and preliminary discussions of the affected morphology provide additional value in the comprehension of the mechanics of physical systems, information which is relevant to *Appendix 1* and *Appendix 2*.

22. Roth, L. et al. 1984. *Giftpflanzen, Pflanzengifte* [Poisonous plants and plant toxins]. Munich: Ecomed.

    Geared towards pharmacists, doctors, and other professionals, this book features toxicological reviews of common herbs in trade and popular ornamental plants. It includes a review of symptoms of poisoning and supportive medical treatment, focusing mainly on human poisoning. The work draws heavily from *Hagers Handbuch*, but reviews other more obscure European texts. As with *Hagers Handbuch*, translation of this German language text was provided by Christopher Hobbs, L.Ac.

23. Tang, W. and G. Eisenbrand. 1992. *Chinese Drugs of Plant Origin.* New York: Springer-Verlag.

    This text focuses predominately on chemical constituents, primary biological activities, and toxicological data. It is one of the most authoritative texts available in English on the pharmacology and chemistry of Chinese medicinal plants.

24. Van Hellemont, J. 1986. *Compendium de Phytotherapie.* Bruxelles, Belgique: Association Pharmaceutique Belge.

    Translation of this French language reference was provided by Daniel Gagnon. The work provides an additional European perspective from a country which allows the marketing of Traditional Herbal Medicines. Published under the auspices of the Belgian Pharmaceutical Association's Scientific Services, the work is designed to facilitate the tasks of pharmacists, including their understanding of potential toxicity and adverse reactions to herbs and herbal products.

25. Watt, J.M. and M.G. Breyer-Brandwijk. 1962. *The Medicinal and Poisonous Plants of Southern and Eastern Africa.* Edinburgh & London: E. & S. Livingstone Ltd.

    This extremely thorough work includes poisonous plants from the African continent not available elsewhere. Numerous plants from other parts of the world, including common house plants and herbs in trade, are also covered. Extensively referenced, specific incidents of human poisoning from plants not reported in other literature are often detailed.

26. Weiss, R.F. 1988. *Herbal Medicine.* Beaconsfield, England: Beaconsfield Publishers Ltd.

    Doctor Weiss first published his *Lehrbuch der Phytotherapie* in 1960. The work identified here is the English translation of the sixth German edition. Written by a medical practitioner and lifelong advocate of herbal medicine to promote the knowledgeable use of herbs in therapeutic settings, the book is a practical guide to a rational herbal practice. Its role as a reference for the *Botanical Safety Handbook* is generally secondary, in that it substantiates issues discussed by other authorities.

27. Wichtl, M. and N.G. Bisset, eds. 1994. *Herbal Drugs and Phytopharmaceuticals.* Stuttgart: Medpharm GmbH Scientific Publishers.

    Beautifully designed, logically organized and thorough in its analysis of the use of each of the 181 herbs which it examines, this is the pre-eminent book on contemporary European herbal medicine. Like Van Hellemont's work in Belgium, the original German edition, *Teedrogen*, was designed to assist the practicing pharmacist in achieving a knowledgeable relationship with herbs in their raw form. Indications, chemical profile, adulterants, toxicity concerns, and the German regulatory status are included for each of the monographs. In addition, this reference provides the primary source of information for **Standard Dose** for those herbs which are included in this work.

28. Wren, R.C. 1988. Revised by E.M. Williamson and F.J. Evans. *Potter's New Cyclopaedia of Botanical Drugs and Preparations.* Essex, England: C.W. Daniel Co. Ltd.

    *Potter's* has been a standard reference text for herbal medicines since its first publication in 1907. The new edition was revised by Drs. Elizabeth Williamson and Fred Evans, pharmacognosists at London University School of Pharmacy. The current volume contains referenced and valuable information on the quantification of primary constituents, medicinal use, toxicology, and dosage, especially in the form of liquid extract preparations.

29. Yeung, H. 1985. *Handbook of Chinese Herbs and Formulas, volume 1.* Los Angeles: Institute of Chinese Medicine.

    As one of the standard Chinese herbal *materia medicas* in English, this volume is directed primarily to practitioners of Traditional Chinese Medicine, and is focused on the traditional use of these herbs. Moderate pharmacology and chemistry data are included for some of the listed plants.

30. AHPA Safety & Labeling Guidelines SubCommittee: Michael McGuffin, Chair; Christopher Hobbs, L.Ac.; and Roy Upton.

    As stated in the *Introduction,* this reference represents consensus of the Committee.

# Bibliography

Abe, F. et al. 1992. Cardiac glycosides from the leaves of *Thevetia neriifolia*. *Phytochemistry. (Oxford)*. 31(9):3189-93.

Abe, F. and T. Yamauchi. 1994. Cardenolide glycosides from the roots of *Apocynum cannabinum*. *Chemical & Pharmaceutical Bulletin*. 42(10):2028-31.

Abel, G. and O. Schimmer. 1983. Induction of structural chromosome aberrations and sister chromatid exchanges in human lymphocytes in vitro by aristolochic acid. *Human Genetics*. 64(2):131-33.

Albuquerque, A.A. et al. 1995. Effects of essential oil of *Croton zehntneri*, and of anethole and estragole on skeletal muscles. *Journal of Ethnopharmacology*. 149(1):41-49.

Ambasta, S.P., ed. 1986. *The Useful Plants of India*. New Delhi: Publications & Information Directorate.

Anderson, I.B. 1996. Pennyroyal toxicity: measurement of toxic metabolite levels in two cases and review of the literature. *Annals of Internal Medicine*. 124(8):726-34.

Anonymous. 1719. *The Compleat Herbal: or the Botanical Institutions of Mr. Tournefort, Chief Botanist to the Late French King. Carefully Translated from the Original Latin*. London: Bonwicke, Goodwin, Walthoe, Wotton, Manship, Wilkin, Tooke, Smith and Ward.

Anonymous. 1987. Ground Coffee. *Consumer Reports*. September. 527-33.

Anonymous. 1988. *Pyrrolizidine Alkaloids*. Geneva: Switzerland: World Health Organization.

Anonymous. 1996. *Briefing Material for Food Products Containing Ephedrine Alkaloids*. Washington: Prepared for the Food Advisory Committee meeting of August 27-28, 1996.

Anthony, A. et al. 1987. Metabolism of estragole in rat and mouse and influence of dose size on excretion of the proximate carcinogen 1'-hydroxyestragole. *Food and Chemical Toxicology*. 25(11):799-806.

Arvigo, R. and M. Balick. 1993. *Rainforest Remedies*. Twin Lakes, WI: Lotus Press.

Awang, D.V.C. 1990. Herbal Medicine: Borage. *Canadian Pharmaceutical Journal*. March 1990:121-26.

Awang, D.V.C., B.A. Dawson, J. Fillion, M. Girard, and D. Kindack. 1993. Echimidine Content of Commercial Comfrey (*Symphytum* spp.-Boraginaceae). *Journal of Herbs, Spices & Medicinal Plants*, vol 2 (1). Binghamton, NY: The Haworth Press.

Bailey, L.H. and E.Z Bailey. 1976. *Hortus Third: A Concise Dictionary of Plants Cultivated in the United States and Canada*. New York: Macmillan Pub. Co.

Baker, C., Chair. 1990. *Report of the South Australian Working Party on Natural and Nutritional Substances*. South Australian Working Party on Natural and Nutritional Substances, South Australian Health Commission.

Batiste - Alentorn, M. et al. 1995. Genotoxic evaluation of ten carcinogens in the Drosophilia melanogaster wing spot test. *Experientia*. 51(1):73-76.

Bensky, D. and A. Gamble. 1986. *Chinese Herbal Medicine.* Seattle: Eastland Press.

Bensky, D. and A. Gamble. 1993. *Chinese Herbal Medicine,* 2nd ed. Seattle: Eastland Press.

Benoni, H. et al. 1996. Studies on the essential oil from guarana. *Zeitschrift fur Lebensmittel-Untersuchung und -Forschung.* 203(1):95-98.

Biddle, J. 1895. *Materia Medica and Therapeutics.* Philadelphia: P. Blakiston's Son & Co.

Bingel, A. and N. Farnsworth. 1980. Botanical Sources of Fertility Regulating Agents: Chemistry and Pharmacology. In *Progress in Hormone Biochemistry and Pharmacology* vol. 1. Eds. Briggs, M. and A. Corbin. St. Albans, VT: Eden Medical Research.

Blackburn, J.L., Chair. 1993. *Second Report of the Expert Advisory Committee on Herbs and Botanical Preparations to the Health Protection Branch, Health Canada.* Ministry of Health Canada.

Blaustein, M.P. 1993. Physiological effects of endogenous ouabain: control of intracellular Ca2+ stores and cell responsiveness. *American Journal of Physiology.* June, 264(6 Pt 1):C1367-87.

Blumenthal, M. 1991. Debunking the "Ginseng Abuse Syndrome." *Whole Foods.* March 1991:89.

Blumenthal, M. 1992. AHPA Petitions FDA for Approval of Stevia Leaf Sweetener. *HerbalGram.* 26:22,55.

Blumenthal, M., ed.; S. Klein, trans. 1993. *Therapeutic Monographs on Medicinal Plants for Human Use of the Commission E Special Expert Committee, Federal Health Agency, Germany* (Draft). Austin: American Botanical Council.

Bombardelli, E. and P. Morazzoni. 1995. *Vitis vinifera* L. *Fitoterapia* 66(4):291-316.

Bone, K. 1995. Juniper Berry is not a kidney irritant. *British Journal of Phytotherapy.* 4:47-48.

Bone, K. 1993/94. Kava - A Safe Herbal Treatment for Anxiety. *British Journal of Phytotherapy.* 3(4):147-53.

Bonokovsky, H.L. et al. 1992. Porphyrogenic properties of the terpenes camphor, pinene, and thujone. *Biochemical Pharmacology.* 43(11):2359-68.

Bradley, P.R., ed. 1992. *British Herbal Compendium, volume 1.* Dorset: British Herbal Medicine Association.

Brody, T. M. 1994. *Human Pharmacology—Molecular to Clinical.* St. Louis: Mosby-Year Book, Inc.

Bruneton, J. 1995. *Pharmacognosy, Phytochemistry, Medicinal Plants.* England: Intercept, Ltd.

Burnham, T.H., ed. 1996. Burdock. *The Review of Natural Products.* Dec., 1996.

Caldwell, J. and J.D. Sutton. 1988. Influence of dose size on the disposition of trans-(methoxy-14C) anethole in human volunteers. *Food and Chemical Toxicity.* 26:87-91.

Carls, N. and R.H. Schiestl. 1994. Evaluation of the yeast DEL assay with 10 compounds selected by the International Program on Chemical Safety for the Evaluation of Short-Term Tests for Carcinogens. *Mutation Research.* 320(4):293-303.

Chadha, Y.R. et al., eds. 1952-88. *The Wealth of India* (Raw Materials), 11 vols. New Delhi: Publications and Information Directorate, CSIR.

Chalker, R. and C. Downer. 1992. *A Woman's Book of Choices: Abortion, Menstrual Extraction, RU-486*. New York: Four Walls Eight Windows.

Chan, V.S. and J. Caldwell. 1992. Comparative induction of unscheduled DNA synthesis in cultured rat hepatocytes by allylbenzenes and their 1'-hydroxy metabolites. *Food and Chemical Toxicology*. 30(10):831-36.

Chan, W. et al. 1983. Clinical observation on the uterotonic effect of I-mu Ts'ao (*Leonurus artemisia*). *American Journal of Chinese Medicine*. 11(1-4):77-83.

Chan, W.Y. 1993. Developmental toxicity and teratogenicity of trichosanthin, a ribosome-inactivating protein, in mice. *Teratogenesis, Carcinogenesis, and Mutagenesis*. 13(2):47-57.

Chang, H.M. and P.P.H. But. 1986. *Pharmacology and Applications of Chinese Materia Medica*. Philadelphia: World Scientific.

Chapman, V.J. 1970. *Seaweeds and their Uses*, 2nd ed. London: Methuen & Co. Ltd.

Chen, S.J. et al. 1995. C-fiber-evoked autonomic cardiovascular effects after inject of Piper beetle inflorescence extracts. *Journal of Ethnopharmacology*. 45(3):183-88.

Chin, R.K. 1991. Ginseng and common pregnancy disorders. *Asia-Oceania Journal of Obstetrics and Gynaecology*. 17(4):379-80.

Christopher, J. 1976. *School of Natural Healing*. Springville: Christopher Publications.

Cook, E.F. and E.W. Martin. 1948. *Remington's Practice of Pharmacy*, 9th ed. Easton, PA: The Mack Publishing Company.

Crellin, J. K. and J. Philpott. 1990. *Herbal Medicine Past and Present; vol. II: A reference guide to medicinal plants*. London: Duke University Press.

Culvenor, C.C.J. et al. 1980. Structure and toxicity of the alkaloids of Russian comfrey (*Symphytum* x *uplandicum* Nyman), a medicinal herb and item of human diet. *Experientia*. 36:377-79.

De Carvalho, J.E. et al. 1991. Cardiac glycosides isolated from the Indian-snuff, *Maquira sclerophyll* Ducke. *Memorias do Instituto Oswaldo Cruz Rio de Janeiro*. 86, n. SUPPL. 2:235-36.

DeFeudis, F.V. 1991. *Ginkgo biloba Extract (EGb 761): Pharmacological Activities and Clinical Applications*. Paris: Elsevier.

Dentali, S.J. 1997. *Herb Safety Review - Kava Piper methysticum Forster f. (Piperaceae)*. Boulder, CO: Herb Research Foundation.

De Smet, P.A.G.M. et al., eds. 1992. *Adverse Effects of Herbal Drugs 1*. New York: Springer-Verlag.

De Smet, P.A.G.M. et al., eds. 1993. *Adverse Effects of Herbal Drugs 2*. New York: Springer-Verlag.

Dragendorff, G. 1898. *Die Heilpflanzen der Verschiedened Völker und Zeiten*. Stuttgart: Verlag von Ferdinand Enke.

Duke, J.A. 1989. *CRC Handbook of Medicinal Herbs*. 7th printing. Boca Raton, FL: CRC Press.

Duke, J.A. 1992. *Handbook of Phytochemical Constituents of GRAS Herbs and Other Economic Plants*. Boca Raton, FL: CRC Press.

Duker, E. et al. 1991. Effects of extracts from *Cimicifuga racemosa* on gonadotropin release in menopausal women and ovariectomized rats. *Planta Medica*. 57(5):420-24.

Dutta, H. et al. 1996. Effect of two cardiac glycosides, digitoxin and digoxin on blood lipids. *Indian Journal of Biochemistry and Biophysics*. Feb. 33(1):76-80.

Ellenhorn, M. and D. Barceloux. 1988. *Medical Toxicology Diagnosis and Treatment of Human Poisoning*. New York: Elsevier.

Ellingwood, F. 1983. *American Materia Medica, Therapeutics and Pharmacognosy*. Vol. II. Naturopathic Medical Series. Portland: Eclectic Medical Publications.

Erdmann, E. 1986. Value of digitalis. An interview with Professor Dr. Erland Erdmann, Munich. *Medizinische Monatsschrift fur Pharmazeuten*. Jan. 9. (2) 47-48.

Erdmann, E. 1995. Digitalis-friend or foe? *European Heart Journal*. 16, n.SUPPL. F:16-19.

Evans, W.C. 1989. *Trease and Evans' Pharmacognosy*, 13th ed. London: Baillière Tindall.

Farnsworth, N. 1985. Siberian Ginseng (*Eleutherococcus senticosus*): Current Status as an Adaptogen. In *Economic and Medicinal Plant Research*, vol. 1. Wagner, H., H. Hikino and N. Farnsworth, eds. London: Academic Press Inc., Ltd.

Farrell, K.T. 1990. *Spices, Condiments and Seasonings*, 2nd ed. New York: Van Nostrand Reinhold.

Federal Register. 1985. 50 FR2124 et seq. (Jan. 15).

Fehr, D. 1982. Bestimmung fluchtiger Inhaltsstoffe in Teezubereitungen. 1. Mitteilung: Freisetzung des atherischen Ols aus Fenchenlfruchten. *Pharmazeutische Zeitung-Nachrichten*. 127:2520-22.

Feldman, E.G. and D. Davidson. 1986. *Handbook of Nonprescription Drugs*. 8th ed. Washington, D.C.: American Pharmaceutical Association.

Felter, H.W. and J.U. Lloyd. 1898. *King's American Dispensatory*. Cincinnati: The Ohio Valley Co.

Fernando, R.C. et al. 1993. Formation and persistence of specific purine DNA adducts by 32P-postlabelling in target and non-target organs of rats treated with aristolochic acid I. Iarc *Scientific Publications*. 124:167-71.

Foster, S., ed. 1992. *Herbs of Commerce*. Austin: American Herbal Products Association.

Friedman, L. et al. 1979. Testicular Atrophy and Impaired Spermatogenesis in Rats Fed High Levels of the Methylxanthines Caffeine, Theobromine, or Theophylline. *Journal of Environmental Pathology and Toxicology*. Jan.-Feb. 2:687-706.

Frohne, D. and H. Pfänder. 1983. *A Colour Atlas of Poisonous Plants*. London: Wolfe Publishing.

Fuller, T.C. and E. McClintock. 1986. *Poisonous Plants of California*. Berkeley: University of California Press.

Furia, T.E. and N. Bellanca. 1971. *Feranoli's Handbook of Flavor Ingredients*. Cleveland, OH: The Chemical Rubber Co.

Gawin, F. 1988. Cocaine and Other Stimulants: Actions, Abuse and Treatment. *New England Journal of Medicine*. 318:1127.

Gheorghiade, M. and B.J. Zarowitz. 1992. Review of randomized trials of digoxin therapy in patients with chronic heart failure. *American Journal of Cardiology.* Jun 4. 69(18):48G-62G; discussion 62G-63G.

Gilbert, R. 1976. Caffeine as a Drug of Abuse. In Gibbins, et al., eds. *Research Advances in Alcohol and Drug Problems.* New York: John Wiley & Sons.

Gilbert, R. 1984. Caffeine Consumption. In G.A. Spiller, ed. *The Methylxanthine beverages and foods: Chemistry, Consumption, and Health Effects.* New York: Liss.

Gilman, A.G. et al. 1985. *Goodman and Gilman's The Pharmacological Basis of Therapeutics,* 7th ed. New York: Macmillan Publishing Company.

Gosselin, R.E. et al. 1984. *Clinical Toxicology of Commercial Products.* Baltimore: Williams & Wilkins.

Graham, H.N. 1992. Green Tea Composition, Consumption, and Polyphenol Chemistry. *Preventive Medicine.* 21:334-50.

Grases, F. et al. 1994. Urolithiasis and phytotherapy. *Internation Urology and Nephrology.* 26(5):507-11.

Graven, E.H. et al. 1992. Antimicrobial and antioxidative properties of the volatile oil of *Artemisia afra* Jacq. *Flavour and Fragrance Journal.* 7(3):121-23.

Greenberg, J.H. et al. 1978. Molar volume relationships and the specific inhibition of a synaptosomal enzyme by psychoactive cannabinoids. *Journal of Medicinal Chemistry.* 21(12):1208-12.

Grieve, M. 1931. *A Modern Herbal.* London: Jonathan Cape.

Gupta, K.P. et al. 1993. Formation and persistence of safrole-DNA adducts over a 10,000-fold dose range in mouse liver. *Carcinogenesis.* 14(8):1517-21.

Harborne, J. and H. Baxter. 1993. *Phytochemical Dictionary.* London: Taylor and Francis.

Hardin, J.W. and J.M. Arena. 1974. *Human Poisoning from Native and Cultivated Plants,* 2nd ed. Durham, NC: Duke University Press.

Hardman, J. and L. Limbird, eds. 1996. *Goodman & Gilman's The Pharmacological Basis of Therapeutics.* 9th ed. New York: McGraw-Hill.

Harkrader, R. et al. 1990. The history, chemistry, and pharmacokinetics of Sanguinaria extract. *Journal of the Canadian Dental Association.* 56(7 Suppl):7-12.

Hasheminejad, G. and J. Caldwell. 1994. Genotoxicity of the alkenylbenzenes alpha- and beta-asarone, myristicin and elimicin as determined by the UDS assay in cultured rat hepatocytes. *Food and Chemical Toxicology.* 32(3):223-31.

Heath, H.B. 1981. *Source Book of Flavors.* New York: Von Nostrand Reinhold.

Heikes, D.L. 1994. SFE with GC and MS determination of safrole and related allylbenzenes in sassafras teas. *Journal of Chromatographic Science.* 32(7):253-58.

Heskel, N. et al. 1983. Phytophotodermatitis due to *Ruta graveolens. Contact Dermatitis.* 9(4):278-80.

Hobbs, C. 1990. *Vitex: The Women's Herb.* Santa Cruz, CA: Botanica Press.

Hobbs, C. 1991a. *Comfrey: A Literature Review.* Santa Cruz, CA: Botanica Press.

Hobbs, C. 1991b. *Ginkgo, Elixir of Youth.* Santa Cruz, CA: Botanica Press.

Hobbs, C. 1996a. *The Ginsengs.* Santa Cruz, CA: Botanica Press.

Hobbs, C. 1996b. *The Herbal Prescriber.* Santa Cruz, CA: Botanica Press.

Hobbs, C. 1997. Vitex, A Literature Review. In press. *Herbalgram*.

Hoffmann, D. 1983. *The Holistic Herbal*. Findhorn, Scotland: The Findhorn Press.

Hoover, J.E., Managing ed. 1970. *Remington's Pharmaceutical Sciences*, 14th ed. Easton, PA: Mack Publishing Company.

Horvat, B. 1993. Galactose-binding lectins as markers of pregnancy-related glycoproteins. *Histochemistry*. 99(1):95-101.

Howes, A.J. et al. 1990. Structure-specificity of the genotoxicity of some naturally occurring alkenylbenzenes determined by the unscheduled DNA synthesis assay in rat hepatocytes. *Food and Chemical Toxicology*. 28(8):537-42.

Hsu, H. 1986. *Oriental Materia Medica, A Concise Guide*. Long Beach: Oriental Healing Arts.

Huang, K.C. 1993. *The Pharmacology of Chinese Herbs*. Boca Raton, FL: CRC Press.

Huff, B.B., ed. 1989. *Physicians' Desk Reference*, 43rd ed. Oradell, NJ: Medical Economics Company.

Hufford, C. et al. 1988. Antifungal activity of *Trillium grandiflorum* constituents. *Journal of Natural Products*. 51(1):94-98.

Imai, K. and K. Nakachi. 1995. Cross sectional study of effects of drinking green tea on cardiovascular and liver diseases. *British Medical Journal*. 310:693-696.

Imre, Z. and T. Yurdun. 1988. Cardiac glycosides from the seeds of *Digitalis cariensis*. *Planta Medica*, 54(6):529-31.

Infante-Rivard, C. et al. 1993. Fetal Loss Associated with Caffeine Intake Before and During Pregnancy. *Journal of the American Medical Association*. 270(24):2940-43.

Ireland, C.M. et al. 1988. Activating mutations in human c-Ha-ras-1 gene induced by reactive derivative of safrole and the glutamic pyrolysis product, Glu-P-3. *Mutagenesis*. 3(5):429-35.

Ishida, T. et al. 1989. Terpenoid biotransformation in mammals. *Xenobiotica*. 19(8):843-55.

Iverson, S.L. et al. 1995. The influence of the p-alkyl substituent on the isomerization of o-quinones to p-quinone methides: potential bioactivation mechanism for catechols. *Chemical Research in Toxicology*. 8(4):537-44.

James, C. et al. 1991. Sesame seed anaphylaxis. *New York State Journal of Medicine*. Oct:457-58.

Jones, K. 1995. *Cat's Claw: Healing Vine of Peru*. Seattle: Sylvan Press.

Kartesz, J.T. 1994. *A Synonymized Checklist of the Vascular Flora of the United States, Canada, and Greenland*, 2nd ed. Portland, OR: Timber Press.

Katcher, B.S. et al. 1983. *Applied Therapeutics - The Clinical Use of Drugs*. 3rd ed. Spokane: Applied Therapeutics, Inc.

Keeler, R. and A. Tu. 1983. *Plant and Fungal Toxins*. New York: Marcel Dekker, Inc.

Kelly, R.A. and T.W. Smith. 1992. Recognition and management of digitalis toxicity. *American Journal of Cardiology*. Jun 4. 69(18):108G-118G; disc.118G-119G.

Kelsey, J. and R. Prevost. 1994. Drug therapy during labor and delivery. *American Journal of Hospital Pharmacy*. 51(19):2394-402.

Keville, K. and M. Green. 1995. *Aromatherapy, A Complete Guide to the Healing Art*. Freedom, CA: The Crossing Press.

Kingsbury, J.M. 1964. *Poisonous Plants of United States and Canada*. Englewood Cliffs, NJ: Prentice-Hall, Inc.

Kingsbury, J.M. 1979. The problem of poisonous plants. In Kinghorn, A.D., ed. *Toxic Plants*. New York: Columbia University Press.

Kluthe, R. et al. 1982. Double blind study of the influence of aristolochic acid on granulocyte phagocytic activity. *Arzneimittel-Forschung*. 32:443-45.

Kong, Y. et al. 1976. Isolation of the uterotonic principle from *Leonurus artemisia*, the Chinese motherwort. *American Journal of Chinese Medicine*. Winter. 4(4):373-82.

Koren, G. et al. 1990. Maternal ginseng use associated with neonatal androgenization. *Journal of the American Medical Association*. 264(22):2866.

Kouzi, S.A. et al. 1994. Hepatotoxicity of germander (*Teucrium chamaedrys* L.) and one of its constituent neoclerodane diterpenes teucrin A in the mouse. *Chemical Research in Toxicology*. 7(6):850-56.

Kruse, J. 1986. Oxytocin: pharmacology and clinical application. *Journal of Family Practice*. 23(5):473-79.

Lampe, K.F. and M.A. McCann. 1985. *AMA Handbook of Poisonous and Injurious Plants*. Chicago: AMA.

Langford, S.D., and P.J. Boor. 1996. Oleander toxicity: an examination of human and animal toxic exposures. *Toxicology*. May 3. 109(1):1-13.

Larrey, D. et al. 1992. Hepatitis after germander (*Teucrium chamaedrys*) administration: another instance of herbal medicine hepatotoxicity. *Annals of Internal Medicine*.117(2):129-32.

Larson, K.M., et al. 1984. Unsaturated pyrrolizidines from borage (*Borago officinalis*), a common garden herb. *Journal of Natural Products*. 47:747-48.

Lebot, V., M. Merlin, and L. Lindstrom. 1992. *Kava: The Pacific Drug*. New Haven: Yale University Press.

Lei, Z-H. et al. 1996. Cardenolides from *Erysimum cheiranthoides*. *Phytochemistry (Oxford)*. 41 (4):1187-89.

Leonard, T.K. et al. 1987. The effects of caffeine on various body systems: a review. *Journal of the American Dietetic Association*. 87(8):1048-53.

Leung, A.Y. 1980. *Encyclopedia of Common Natural Ingredients Used in Food, Drugs and Cosmetics*. New York: John Wiley & Sons.

Leung, A.Y. and S. Foster. 1996. *Encyclopedia of Common Natural Ingredients Used in Food, Drugs and Cosmetics*, 2nd ed. New York: John Wiley & Sons.

Lewin, L. 1931. *Phantastica - Narcotic and Stimulating Drugs: Their Use and Abuse*. London: Kegan Paul, Trench, Trubner & Co., Ltd.

Lewin, L. 1962. *Gifte und Vergiftungen. Lehrbuch der Toxikologie*, 5th ed. Ulm/Donau: Karl F. Haug Verlag.

Lewis, W.H. and M.P.F. Elvin-Lewis. 1977. *Medical Botany. Plants Affecting Man's Health*. New York: John Wiley & Sons.

Linsley, P.I. 1995. Automatic Detention of Stevia Leaves, Extract of Stevia Leaves, and Foods Containing Stevia. FDA Import Alert #45-06. Rev. 9-18.

List, P.H. and L. Hörhammer. 1973-'79. *Hagers Handbuch der Pharmazeut-ischen Praxis*, 7 vols. New York: Springer-Verlag.

Lloyd, J.U. and C.G. Lloyd. 1931. (1886-87). *Drugs and Medicines of North America*, Vol II. Cincinnatti: Lloyd Library.

Longerich, L, et al. 1993. Digoxin-Like Factors in Herbal Teas. *Clinical and Investigative Medicine*, 16, 3:210-18.

Lu, L.J. et al. 1986. 32P-postlabeling assay in mice of transplacental DNA damage induced by the environmental carcinogens safrole, 4-aminobiphenyl, and benzo(a)pyrene. *Cancer Research.* 46(6):3046-54.

Luo, G. et al. 1992. Hydrolysis of the 2', 3'-allylic epoxides of allylbenzene, estragole, eugenol, and safrole by both microsomal and cytosolic epoxide hydrolases. *Drug Metabolisms and Disposition.* 20(3):440-45.

Luo, G. and T.M. Guenthner. 1995. Metabolisms of allylbensene 2', 3' -oxide and estragole 2', 3' -oxide in the isolated perfused rat liver. *Journal of Pharmacology and Experimental Therapeutics.* 272(2):588-96.

Lust, J. 1974. *The Herb Book.* New York: Benedict Lust Publications.

Madaus, G. 1976 (1938). *Lehrbuch der Biologishchen Heilmittel*, 3 vols. New York: Georg Olms Verlag.

Mahato, S. B. et al. 1989. Cardiac glycosides from *Corchorus olitorius. Journal of the Chemical Society Perkin Transactions* I. 11:2065-68.

Maier, P. et al. 1987. Low oxygen tension, as found in tissues *in vivo*, alters the mutagenic activity of aristolochic acid I and II in primary fibroblasts-like rat cells *in vitro. Environmental and Molecular Mutagenesis.* 10(3):275-84.

Makarevich, I F. 1989. Cardiac glycosides from *Cheiranthus allioni:* XIII. Glucoerycordin. *Khimiya Prirodnykh Soedinenii.* 1: 73-75.

Makarevich, I.F. et al. 1991. Cardiac glycosides from *Erysimum contractum. Khimiya Prirodnykh Soedinenii.* 1:58-62.

Malinow, M.R. et al. 1982. Systemic lupus erythematosus-like syndrome in monkeys fed alfalfa sprouts: Role of a nonprotein amino acid. *Science.* 216:415.

Maraganore, J.M. et al. 1987. Purification and characterization of trichosanthin. Homology to the ricin A chain and implications as to mechanism of abortifacient activity. *Journal of Biological Chemistry.* 262(24):11628-33.

Martin, R A. et al. 1991. Cardenolides from *Asclepias asperula* ssp. *capricornu* and *Asclepias viridis. Phytochemistry (Oxford).* 30(12):3935-40.

Matsuda, H. 1991. Pharmacological studies on leaf of *Arctostaphylos uva-ursi* (L.) Combined effect of arbutin and indomethacin on immuno-inflammation. *Yakugaku Zasshi. Journal of the Pharmaceutical Society of Japan.* 111(4-5):253-58.

Mayer, R. and M. Mayer. 1949. Biologische salicyltherapie mit *Cortex Salicis* (Weidenrinde). *Pharmazie* 4: 77-81.

McKenzie, R.A. et al. 1992. Suspected poisoning of cattle. *Australian Veterinary Journal.* May. 69(5):117-18.

Mcvann, A. et al. 1992. Cardiac glycoside poisoning involved in deaths from traditional medicines. *South African Medical Journal.* 81(3):139-41.

Medford R.M. 1993. Digitalis and the Na+,K(+)-ATPase. *Heart Disease and Stroke.* May-Jun. 2(3):250-55.

Mengs, I. 1988. Tumor induction in mice following exposure to aristolochic acid. *Archive in Toxicology.* 61(6):504-5.

Mengs, U. 1987. Acute toxicity of aristolochic acid in rodents. *Archives of Toxicology.* 59:328-31.

Mengs, U. and M. Klein. 1988. Genotoxic effects of aristolochic acid in the mouse micronucleus test. *Planta Medica.* 54:502-3.
Merck & Co. 1930. *Merck's Index*, 4th ed. Rahway, NJ: Merck & Co. Inc.
Merck & Co. 1989. *Merck's Index*, 11th ed. Rahway, NJ: Merck & Co. Inc.
Michols, D.M. 1995. Letter to Nonprescription Drugs Manufacturers Association of Canada on letterhead of Health Protection Branch, Health Canada, dated September 22, 1995, and Attachment.
Miller, J.A. et al. 1982. The metabolic activation and carcinogenicity of alkenylbenzenes that occur naturally in many spices. In: Stich, H.F., ed. *Carcinogens and mutagens in the environment.* Vol I. Boca Raton, FL: CRC Press.
Miller, J.A. and E.C. Miller. 1983. The metabolic activation and nucleic acid adducts of naturally-occurring carcinogens: recent results with ethyl carbamate and the spice flavors safrole and estragole. *British Journal of Cancer.* 48(1):1-15.
Millet, Y. et al. 1981. Toxicity of some essential plant oils. Clinical and experimental study. *Clinical Toxicology.* 18(12):1485-98.
Milspaugh, C.F. 1887. *American Medicinal Plants.* New York: Boericke & Tafel.
Mitchell, H.W. et al. 1983. *British Herbal Pharmacopoeia*, 4th impression, 1991. Bournemouth, U.K.: British Herbal Medicine Association.
Mitchell, J.C. and A. Rook. 1979. *Botanical Dermatology.* Vancouver: Greenglass, Ltd.
Moffet, H.H. 1995. Acupuncture and Oriental Medicine Update. *Alternative & Complementary Therapies.* 1(3):193.
Moore, M. 1979. *Medicinal Plants of the Mountain West.* Santa Fe: Museum of New Mexico Press.
Moore, M. 1990. *Los Remedios*, 2nd ed. Santa Fe: Red Crane Books.
Moore, M. 1993. *Medicinal Plants of the Pacific West.* Santa Fe: Red Crane Books.
Morrazzoni, P. and E. Bombardelli. 1995. *Valeriana officinalis*: traditional use and recent evaluation of activity. *Fitoterapia.* 66(2):99-112.
Moreno, J.J. 1993. Effect of aristolochic acid on arachidonic acid cascade and *in vivo* models of inflammation. *Immunopharmacology.* 26(1):1-9.
Mose, J.R. et al. 1980. Effect of aristolochic acid on herpes simplex infection of the rabbit eye. *Arzneimittel-Forschung* 30:1571-73.
Muenscher, W.C. 1951. *Poisonous Plants of the United States.* New York: Macmillan Company.
Muller, L. et al. 1994. The genotoxic potential *in vitro* and *in vivo* of the allyl benzene etheric oils estragole, basil oil and trans-anethole. *Mutation Research.* 325(4):129-36.
Nagasawa, H. et al. 1992. Further study on the effects of motherwort (*Leonurus sibiricus* L) on preneoplastic and neoplastic mammary gland growth in multiparous GR/A mice. *Anticancer Research.* Jan-Feb. 12(1):141-43.
Narod, S. et al. 1991. Coffee during pregnancy: a reproductive hazard? *American Journal of Obstetrics and Gynecology.* Apr. 164(4):1109-14.
Office of the Federal Register, publ. 1994. *Code of Federal Regulations, Title 21: Food and Drugs.* Washington, D.C.: U.S. Government Printing Office.
Okunade, A.L. et al. 1987. Estragole: an acute toxic prinicple from the volatile oil of the leaves of *Clausena anisata. Journal of Natural Products.* 50(5):990-91.

Olayiwola, A. 1992. WHO Guidelines for the assessment of herbal medicines. *Fitoterapia.* LXIII(2):99.

Osol, A. and G.E. Farrar. 1947. *The Dispensatory of the United States of America,* 24th ed. Philadelphia: J.B. Lippincott Company.

Osol, A. and G.E. Farrar. 1955. *The Dispensatory of the United States of America,* 25th ed. Philadelphia: J.B. Lippincott Company.

Pammel, L.H. 1910. *A Manual of Poisonous Plants.* Cedar Rapids: The Torch Press.

Penna, M. 1946. *Dicionario Brasileiro des Plantes Medicinais.* Rio de Janiero: Eric Eichner and CIA. LTDA.

Perry, L.M. 1980. *Medicinal Plants of East and Southeast Asia.* Cambridge: The MIT Press.

Pezzuto, J.M. et al. 1986. Evaluation of the mutagenic and cytostatic potential of aristolochic acid (3,4-methylenedioxy-8-methoxy-10-nitrophenanthrene-1-carboxylic acid) and several of its derivatives. *Mutation Research.* 206(4):447-54.

Pfau, W. et al. 1990. Aristolochic acid binds covalently to the exocyclic amino group of purine nucleotides in DNA. *Carcinogensis.* 11(2):313-19.

Pillion, D. et al. 1996. Structure-function relationship among Quillaja saponins serving as excipients for nasal and ocular delivery of insulin. *Journal of Pharmaceutical Sciences.* May. 85(5):518-24.

Pizzorno, J.E. and M.T. Murray. 1992. *A Textbook of Natural Medicine.* Seattle: John Bastyr Publications.

Powers, J.L., E.H. Wirth, A.B. Nichols, et al. 1942. *The National Formulary,* 7th ed. Washington, D.C.: American Pharmaceutical Association.

Prakash, K. et al. 1991. Two pregnane glycosides from *Hemidesmus indicus. Phytochemistry (Oxford).* 30(1). 297-300.

Qato, M.K. and T.M. Guenthner. 1995. 32P-postlabeling analysis of adducts formed between DNA and safrole 2', 3'-epoxide: absence of adduct formation *in vivo. Toxicology Letters.* 75(1-3):201-7.

Rakel, R. ed. 1996. *Conn's Current Therapy.* Philadelphia: W.B. Saunders Co.

Randerath, K. 1993. Altered fidelity of a nucleic acid modifying enzyme, T4 polynucleotide kinase, by safrole-induced DNA damage. *Carcinogensis.* 14(8):1523-9.

Randerath, K. et al. 1993. Flavor constituents in cola drinks induce hepatic DNA adducts in adult and fetal mice. *Biochemical and Biophysical Research Communications.* 192(1):61-68.

Rasheed, R.A. et al. 1995. Effect of *Teucrium stocksianum* on paracetamol-induced hepatotoxicity in mice. *General Pharmacology.* 26:297-301.

Regoli, D. 1986. Kinins, receptors, antagonists. *Advances in Experimental Medicine and Biology.* 198 Pt A:549-58.

Remington, J.P. and H.C. Wood. 1918. *The Dispensatory of the United States of America,* 20th ed. Philadelphia: J.B. Lippincott Company.

Reynolds, J.E., ed. 1989. *Martindale: The Extra Pharmacopoeia,* 29th ed. London: The Pharmaceutical Press.

Reynolds, J.E., ed. 1993. *Martindale: The Extra Pharmacopoeia,* 30th ed. London: The Pharmaceutical Press.

Riddle, J.M. 1985. *Dioscorides on Pharmacy and Medicine.* Austin: University of Texas Press.

Roitman, J.N. 1983. Ingestion of Pyrrolizidine Alkaloids: A Health Hazard of Global Proportions. *Xenobiotics in Foods and Fields.* J.W. Finley, ed. ACS Symposium Series 234. Washington, D.C.: American Chemical Society.

Ronnberg, B. et al. 1995. Adjuvant activity of non-toxic *Quillaja saponaria* Molina components for use in ISCOM matrix. *Vaccine.* 13(14):1375-82.

Rosenthal, M.D. et al. 1992. The effects of the phospholipase A2 inhibitors aristolochic acid and PGBx on A23187-stimulated mobilization of arachidonate in human neutrophils are overcome by diacylglycerol or phorbol ester. *Biochimica et Biophysica Act.* 1126(3):319-26.

Ross, M. and K. Brain. 1977. *An Introduction to Phytopharmacy.* Bath, U.K.: Pitman Medical Press.

Roth, L. et al. 1984. *Giftpflanzen, Pflanzengifte.* Munich: Ecomed.

Rubio-Poo, C. et al. 1991. The anticoagulant effect of beta-asarone in the mouse and the rat. *Proceedings of the Western Pharmacology Society.* 34:107-12.

Sangster, S.A. et al. 1987. The metabolic diposition of [methoxy-14C]-labelled trans-anethole, estragole and p-propylanisole in human volunteers. *Xenobiotica.* 17(10):1223-32.

Sato, T. et al. 1993. Digitalis for the treatment in patients with heart failure. *Nippon Rinsho. Japanese Journal of Clinical Medicine.* May. 51(5):1260-67.

Sauer, G.C. 1973. *Manual of Skin Diseases.* Philadelphia: J.B. Lippincott Company.

Schmeiser, H.H. et al. 1996. Detection of DNA adducts formed by aristolochic acid in renal tissue from patients with Chinese herbs nephropathy. *Cancer Research.* 56(9):2025-28.

Schultes, R.E. and R.F. Raffauf. 1990. *The Healing Forest.* Portland, OR: Dioscorides Press.

Schwartz, H. 1986. Safety profile of sanguinarine and sanguinaria extract. *Compendium of Continuing Education in Dentistry.* Suppl 7L:212-17.

Selavka, C. M. 1991. Poppy seed ingestion as a contributing factor to opiate-positive urinalysis results: the Pacific perspective. *Journal of Forensic Sciences.* 36: 685-96.

Shi, M. et al. 1995. Stimulating action of *Carthamus tinctorius* L., *Angelica sinensis* (Oliv.)Diels and *Leonurus sibiricus* L. on the uterus. *Chung-Kuo Chung Yao Tsa Chih China Journal of Chinese Materia Medica.* Mar. 20(3):173-75, 192.

Siegel, R.K. 1979. Ginseng abuse syndrome: problems with the panacea. *Journal of the American Medical Association.* 241:1614-15.

Siering, H. and H. Muller. 1981. Antagonistic effects of glucocorticoids and aristolochic acid on the immunocyte adherence phenomenon. *Arzneimittel-Forschung.* 31(8):1260-62.

Siering, H. and R. Przesang. 1984. Modification of antibody formation in animal experiments by glucocorticoids and aristolochic acid]. *Zeitschrift fur die Gesamte Innere Medizin und Ihre Grenzgebiete.* 39:81-84.

Slawson, M. et al. 1996. Correlations of the induction of microsomal epoxide hydrolase activity with phase II drug conjugating enzyme activities in rat liver. *Toxicology Letters.* 85(2):29-34.

Solis-Cohen, S. and T. Githens. 1928. *Pharmacotherapeutics Materia Medica and Drug Action.* New York: D. Appleton & Co.

Stanton, C. and R. Gray. 1995. Effects of Caffeine Consumption on Delayed Conception. *American Journal of Epidemiology.* 142(12):1322-29.

Stavric, B. 1992. An update on research with coffee/caffeine 1989-1990. *Food and Chemical Toxicology.* 30:533-55.

Suatengco, R. et al. 1995. The significance and work-up of minor gastrointestinal bleeding in hospitalized nursing home patients. *Journal of the National Medical Association.* 87(10):749-50.

Tang, W. and G. Eisenbrand. 1992. *Chinese Drugs of Plant Origin.* New York: Springer-Verlag.

Tippo, O. and W.L. Stern. 1977. *Humanistic Botany.* New York: W.W. Norton & Company.

Tisserand, R.B. 1977. *The Art of Aromatherapy.* Rochester, VT: Healing Arts Press.

Trease, G.E. and W.C. Evans. 1978. *Pharmacognosy.* 11th edition. Bailliere Tindall: London.

Tsai, R.S. et al. 1994. Structure-genotoxicity relationships of allylbenzenes and propenylbenzenes: a quantum chemical study. *Chemical Research in Toxicology.* 8(1):164.

Tsao, S.W. et al. 1986. Selective killing of choriocarcinoma cells *in vitro* by trichoasnthin, a plant protein purified from root tubers of the Chinese medicinal herb *Trichosanthes kirilowii. Toxicon.* 24(8):831-40.

Tsao, S.W. et al. 1990. Toxicities of trichosanthin and alpha-momorcharin, abortifacient proteins from Chinese medicinal plants, on cultured tumor cell lines. *Toxicon.* 28(10):1183-92.

Tu, G., ed. 1988. *Pharmacopoeia of the People's Republic of China (English Edition 1988).* Beijing: People's Medical Publishing House.

Turner, N.J. and A.F. Szazawinski. 1991. *Common Poisonous Plants and Mushrooms of North America.* Portland, OR: Timber Press, Inc.

Truitt, E.B. et al. 1963. Evidence of monoamine oxidase (MAO) inhibition by myristicin and nutmeg. *Proceedings of the Society of Experimental Biology and Medicine.* 112:647-50.

Tyler, V. 1993. *The Honest Herbal.* 3rd ed. Binghamton, NY: Pharmaceutical Products Press.

Tyler, V. 1994. *Herbs of Choice.* Binghamton, NY: Pharmaceutical Products Press.

United States Pharmacopoeia Convention. 1820. *United States Pharmacopoeia,* 1st ed. Boston: Wells & Lilly.

United States Pharmacopoeia Convention. 1974. *United States Pharmacopoeia,* 19th ed. Rockville, MD: United States Pharmacopoeia Convention.

United States Pharmacopoeial Convention. 1926. *The Pharmacopoeia of the United States of America.* Philadelphia: J.B. Lippincott Company.

Upton, R. et al. 1996. Willow Monograph: *American Herbal Pharmacopoeia.* (unpublished).

Vanhaelen, M. et al. 1994. Identification of aristolochic acid in Chinese herbs [letter]. *Lancet.* Jan 15, 343(8890):174.

Van Hellemont, J. 1986. *Compendium de Phytotherapie.* Bruxelles, Belgique: Association Pharmaceutique Belge.

Vanderweghem, J.L. 1994. A new form of nephropathy secondary to the absorption of Chinese herbs. *Bulletin et Memoires de l'Academie Royale de Medicine de Belgique.* 149:128-40.

van Zwieten, P.A. 1994. Pharmacotherapy of congestive heart failure. Currently used and experimental drugs. *Pharmacy World and Science.* Dec 2. 16(6):234-42.

Wagner, H. 1984. *Plant Drug Analysis.* Berlin: Springer-Verlag.

Watt, J.M. and M.G. Breyer-Brandwijk. 1962. *The Medicinal and Poisonous Plants of Southern and Eastern Africa.* Edinburgh & London: E. & S. Livingstone Ltd.

Watts, C. et al. 1994. Letter to Carol Dillard, Executive Director, American Herbal Products Association, June 21, 1994. (unpublished).

Weiss, R.F. 1988. *Herbal Medicine.* Beaconsfield, England: Beaconsfield Publishers Ltd.

Weissmann, G. 1991. Aspirin. *Scientific American.* January 1991:84-90.

Welland, D. 1996. As caffeine controversy rages on, what's a coffee lover to do? *Environmental Nutrition.* 19 (1).

Welsh, F.W. 1995. Letter to Canadian Health Food Association on letterhead of Health Protection Branch, Health Canada, dated February 6, 1995, and Enclosure.

Werbach, M.R. and M.T. Murray. 1994. *Botanical Influences on Illness.* Tarzana, California: Third Line Press.

Wey, W.C. 1820. *United States Pharmacopoeia.* 1st ed. Boston: Wells and Lilly.

Whitmore, A. 1997. FDA warns consumers against dietary supplement products that may contain Digitalis mislabeled as "Plantain". Washington, DC: Press Office; Food and Drug Administration; U.S. Department of Health and Human Services.

Wichtl, M. and N.G. Bisset, eds. 1994. *Herbal Drugs and Phytopharmaceuticals.* Stuttgart: Medpharm GmbH Scientific Publishers.

Windholz, M. et al., eds. 1983. *The Merck Index.* Rahway, NJ: Merck & Co., Inc.

Wiseman, R. et al. 1987. Structure-activity studies of the epatocarcinogenicities of alkenylbenzene derivatives related to estragole and safrole on administration to preweanling male C57BL/6J x C3H/HeJ F1 mice. *Cancer Research.* 47(9):2275-83.

Wood, H.C. and A. Osol. 1943. *The Dispensatory of the United States of America,* 23rd ed. Philadelphia: J.B. Lippincott Company.

Wrba, H. et al. 1992. Carcinogenicity testing of some constituents of black pepper (*Piper nigrum*). *Experimental and Toxicologic Pathology.* 44(2):61-65.

Wren, R.C. 1988. Revised by E.M. Williamson and F.J. Evans. *Potter's New Cyclopaedia of Botanical Drugs and Preparations.* Essex, England: C.W. Daniel Co. Ltd.

Xu, H. et al. 1992. Proliferation of lymphocytes T and B by prehispanolone LC-5504 of *Leonurus hereterophyllus* Sweet. *Yao Hsueh Hsueh Pao Acta Pharmaceutica Sinica.* 27(11):812-16.

Yen, K.Y. 1992. *The Illustrated Chinese Materia Medica.* Taipei: SMC Publishing Inc.

Yeung, H. 1985. *Handbook of Chinese Herbs and Formulas, volume 1*. Los Angeles: Institute of Chinese Medicine.

Yeung, H. 1996. *Handbook of Chinese Herbs*. Rosemead, CA: The Institute of Chinese Medicine.

Youngken, H.W. 1921. *A Textbook of Pharmacognosy*. Philadelphia: Blakiston's Son & Co.

Zhang, H. et al. 1994. Protective effects of N-acetyl-L-cysteine in endotoxemia. *American Journal of Physiology*. 299(5 Pt 2):H1746-54.

Zhang, Z.X., et al. 1988. Atratosides A, B, C, and D, steroid glycosides from the root of Cynanchum atratum. *Phytochemistry. (Oxford)* 27(9). 2935-42.

Zheng, Y.T. et al. 1995. *In vitro* immunotoxicity and cytotoxicity of trichosanthin against human normal immunocytes and leukemia-lympoma cells. *Immunopharmacology and Immunotoxicology*. 17(1):69-79.

Zhou, J. Bioactive Glycosides from Chinese Medicines. *Memorias do Instituto Oswaldo Cruz*. 86, Suppl. 2:231-34.

Zimmerman, D.R. 1983. *The Essential Guide to Nonprescription Drugs*. New York: Harper & Row, Publishers, Inc.

# Plant Index

*Abies balsamea*, 3
absinthe, 15, 158
*Achillea millefolium*, 3, 171, 185
achyranthes, 3, 171, 183, 185
*Achyranthes bidentata*, 3, 171, 185
aconite, 3
*Aconitum carmichaelii*, 3, 189
acorus, 3-4, 134-136, 185, 189
*Acorus calamus*, 3-4, 135-136, 185, 189
*Acorus gramineus*, 4, 136, 185, 189
adenophora, 4
*Adenophora stricta*, 4
*Adenophora tetraphylla*, 4
*Adiantum pedatum*, 4, 185
agar, 55-56, 165
agarweed, 55-56
agastache, 5
*Agastache rugosa*, 5
*Agathosma betulina*, 19
*Agathosma crenulata*, 19
*Agrimonia eupatoria*, 5, 157
agrimony, 5
*Agropyron repens*, 45
alant, 64
*Albizia julibrissin*, 5, 157, 171, 185
*Alcea rosea*, 5
*Alchemilla xanthochlora*, 5
alder buckthorn, 5, 95
aletris, 5-6
*Aletris farinosa*, 5-6
alfalfa, 74
alisma, 6
*Alisma orientale*, 6
alkanet, 6, 182-183, 188
*Alkanna tinctoria*, 6, 151, 182, 185, 188
all-heal, 91
*Allium sativum*, 6, 188
*Allium schoenoprasum*, 7
allspice, 86
*Aloe ferox*, 7-8, 179, 185, 188
*Aloe perryi*, 7, 179, 185, 188
*Aloe vera*, 7, 178-179, 183, 185, 188

aloe vera, 7, 178-179, 183, 185, 188
aloe yucca, 124
aloes, 7-8, 188
*Aloysia triphylla*, 8
alpine strawberry, 53
*Alpinia galanga*, 8
*Alpinia officinarum*, 8
*Althaea officinalis*, 9
alumroot, 57, 61
ambal, 85
American chestnut, 24
American elder, 102
American ephedra, 46
American ginseng, 82
American hellebore, 120, 189
American liverleaf, 61, 183
American mandrake, 89
American mistletoe, 85, 189
American pennyroyal, 60, 183
American white hellebore, 120
*Amomum cardamomum*, 44
*Amomum melegueta*, 9
*Amomum tsao-ko*, 9
amur cork tree, 84
*Anaphalis margaritacea*, 9
andrographis, 9, 165, 183, 185
*Andrographis paniculata*, 9, 165, 185
anemarrhena, 9
*Anemarrhena asphodeloides*, 9
*Anemopsis californica*, 10
*Anethum graveolens*, 10
angelica, 10-11, 171, 176-177, 183, 185
*Angelica archangelica*, 10, 171, 177, 185
*Angelica atropurpurea*, 10, 171, 177, 185
*Angelica dahurica*, 10
*Angelica pubescens*, 10, 177
*Angelica sinensis*, 11, 185
*Aniba rasaeodora*, 11
anise, 63, 86, 183
annatto, 20
*Anthriscus cerefolium*, 11, 185

*Apium graveolens*, 11, 177, 185
*Apocynum androsaemifolium*, 12, 141, 189
*Apocynum cannabinum*, 12
apricot, 91-92, 141-142, 189
*Aralia californica*, 12, 185
*Aralia nudicaulis*, 12, 185
*Aralia racemosa*, 12, 185
*Arctium lappa*, 12
*Arctostaphylos uva-ursi*, 13, 157, 172
*Arisaema japonicum*, 13, 185
*Arisaema triphyllum*, 13
*Aristolochia clematitis*, 13-14, 132, 185, 189
*Aristolochia contorta*, 14, 132, 185, 189
*Aristolochia debilis*, 14, 132, 185, 189
*Aristolochia fangchi*, 131
*Aristolochia serpentaria*, 14, 132, 185, 189
*Armoracia rusticana*, 14
arnica, 14, 183, 185, 189
*Arnica latifolia*, 14, 185, 189
*Arnica montana*, 14, 189
aromatic Solomon's seal, 89
arrowroot, 74
*Artemisia abrotanum*, 15, 171, 185
*Artemisia absinthium*, 15, 158-159, 171, 185, 188
*Artemisia annua*, 15, 185
*Artemisia capillaris*, 15-16, 185
*Artemisia douglasiana*, 16, 171, 185
*Artemisia dracunculus*, 16, 144-145
*Artemisia lactiflora*, 16, 171, 185
*Artemisia scoparia*, 15, 185
*Artemisia vulgaris*, 16, 171, 185
asafetida, 52, 169, 183
asafoetida, 52
*asarabacca*, 16, 183
*Asarum canadense*, 16, 132, 136, 171, 185
*Asarum europaeum*, 16, 136, 169, 185
*Asclepias asperula*, 17, 185
*Asclepias tuberosa*, 17, 141, 169, 185
ashwagandha, 124, 183
Asiatic dogwood, 37
asparagus, 17
*Asparagus cochinchinensis*, 17
*Asparagus officinalis*, 17

*Asperula odorata*, 54
astragalus, 17
*Astragalus membranaceus*, 17
*Astragalus mongholicus*, 17
Atlantic yam, 42
atractylodes, 18
*Atractylodes lancea*, 18
*Atractylodes macrocephala*, 18
*Atropa belladonna*, 18, 133-134, 189
*Avena fatua*, 18
*Avena sativa*, 18

*ba ji tian*, 77
*bai dou kou*, 44
*bai he*, 70
*bai qu cai*, 28
*bai shao*, 80
*bai wei*, 40
*bai zhi*, 10
*bai zhu*, 18
*bai zi ren*, 88
Baikal skullcap, 105
balloon flower, 88
*Ballota nigra*, 18
balm, 3, 75
balm of Gilead, 3, 90
balmony, 28
balsam fir, 3
*ban lan gen*, 64
*ban xia*, 86
*Baptisia tinctoria*, 18, 185
barberry, 19-20, 73, 183
barley, 61, 183
*Barosma betulina*, 19, 185
*Barosma crenulata*, 19, 185
*Barosma serratifolia*, 19, 185
barrenwort, 47
basil, 79, 144, 183, 188
*Bauhinia forficata*, 19
bay leaf, 68
bayberry, 77
beach silvertop, 58
beaked filbert, 37
beaked hazel, 37
bearberry, 13
beard moss, 119
beebalm, 76, 183

INDEX

behada, 114
*bei mu*, 54
*bei sha shen*, 58
belladonna, 12, 18, 105, 133-134, 189
Belleric myrobalan, 114
bellflower, 34
*Benincasa hispida*, 19
benzoin, 111
*Berberis aquifolium*, 73
*Berberis vulgaris*, 19-20, 138, 185
bergamot orange, 33
beth root, 117, 164, 183
*Betula lenta*, 20, 155
*Betula pendula*, 20, 155
*Betula pubescens*, 20, 155
bilberry, 119
biota, 88
*Biota orientalis*, 88
birth root, 117
bissy nut, 35
bistort, 90
bitter almond, 92, 141, 189
bitter orange, 32
bitter root, 12
*Bixa orellana*, 20
black birch, 20
black cohosh, 29-30, 183, 188
black currant, 98
black haw, 122
black horehound, 18
black mustard, 21
black pepper, 87
black peppercorns, 87
black root, 69
black snakeroot, 29
black tea, 22
black walnut, 65
blackberry, 99
bladderwrack, 54, 146, 183, 188
blazing star, 5
blessed thistle, 33, 183
bloodroot, 103, 136-137, 183
blue cohosh, 24, 183
blue flag, 64, 183
blue gum, 49
blue lobelia, 71, 183
blue vervain, 121, 183
blueberry, 119

bogbean, 76
boldo, 84
boneset, 50
borage, 20-21, 150-151, 182-183, 188
*Borago officinalis*, 20-21, 151, 182, 185, 188
*Boswellia carteri*, 21
*Boswellia serrata*, 21
Bourbon vanilla, 120
box holly, 100
boxwood, 22, 189
*Brassica juncea*, 21-22
*Brassica nigra*, 21
Brazil peppertree, 104
Brazilian rhatany, 66
Brazilian rosewood, 11
Brigham tea, 46
British elecampane, 63
broad-leaved dock, 100
broom, 41, 56, 100, 183, 189
broom flower, 56
broom tops, 41
broomrape, 32
Brown's Lily, 70
*bu gu zhi*, 38
*bu za ye*, 76
buchu, 19, 183
buckbean, 76
buckthorn bark, 95
bugbane, 29
bugleweed, 72, 183, 188
*Bupleurum chinense*, 22
*Bupleurum falcatum*, 22
*Bupleurum scorzoneraefolium*, 22
burdock, 12
butcher's broom, 100
butternut, 65
*Buxus sempervirens*, 13, 22, 189


calamus, 3-4, 135-136, 185, 189
calendula, 22
*Calendula officinalis*, 22
California chia, 101
California poppy, 49, 183
California spikenard, 12, 183
calisaya bark, 30
*Calluna vulgaris*, 22

*Camellia sinensis*, 22, 157, 176
camphor, 30-31, 152-153, 158, 183
Canada balsam, 3
Canada-fleabane, 36
Canada snakeroot, 16, 132, 183
canaigre, 100
*Cananga odorata*, 23
*Canarium album*, 23
*cang zhu*, 18
canker-root, 36
*cao guo*, 9
cape aloe, 7
capillary artemisia, 15, 183
*Capsella bursa-pastoris*, 23, 149, 171, 185
*Capsicum annuum*, 23
caraway, 24
cardamom, 44
*Carica papaya*, 24
carob, 26
carrot, 41, 184
*Carthamus tinctorius*, 24, 165, 170-171, 185
*Carum carvi*, 24
cascara sagrada, 96, 178, 183, 188
cassia, 31-32, 106, 178-179, 183, 185
cassia cinnamon, 31
cassia lignea, 31
*Cassia obtusifolia*, 106
*Cassia tora*, 106
cassis, 98
*Castanea dentata*, 24
*Castanea sativa*, 24, 157
castor, 98-99, 147-148, 178-179, 183
Catalonian jasmine, 65
*Catharanthus roseus*, 24, 165, 189
catmint, 79
catnip, 79, 183
cat's claw, 118
catuaba, 49
*Caulophyllum thalictroides*, 24-25, 165, 171, 185
cayenne, 23
*ce bai ye*, 88
*Ceanothus americanus*, 25
cedarwood, 66
celandine, 28, 183
celery, 11, 183

*Centaurea cyanus*, 25
*Centaurium erythraea*, 25
*Centaurium umbellatum*, 25
centaury, 25
*Centella asiatica*, 26
*Cephaelis ipecacuanha*, 26, 168-169, 185
*Ceratonia siliqua*, 26
*Cetraria islandica*, 26, 172
Ceylon cinnamon, 31
cha-de-bugre, 36
*Chaenomeles speciosa*, 27
*chai hu*, 22
*Chamaelirium luteum*, 27, 171-172, 185
*Chamaemelum nobile*, 27, 165, 169, 171, 185
*Chamomilla suaveolens*, 27
*Changium smyrnoides*, 28, 185
chaparral, 67
chapeau de couro, 44
chaste tree, 123, 183
cheese rennet, 54
*Chelidonium majus*, 28, 138, 185
*Chelone glabra*, 28
*chen pi*, 33
cherry birch, 20
chervil, 11, 183
chia, 101
chickweed, 110
chicory, 29
*Chimaphila umbellata*, 28
Chinatree, 74, 189
Chinese asparagus, 17
Chinese clematis, 33
Chinese corktree, 84
Chinese cucumber, 116
Chinese ephedra, 45
Chinese foxglove, 95
Chinese ginseng, 81
Chinese goldthread, 36, 183
Chinese hawthorn, 37
Chinese jointfir, 45
Chinese jujube, 125
Chinese knotweed, 90
Chinese motherwort, 69, 183
Chinese parsley, 36
Chinese rhubarb, 97, 183, 188
Chinese snakegourd, 116

PLANT INDEX

Chinese tea, 22
Chinese thoroughwax, 22
Chinese yam, 42
*Chionanthus virginicus*, 29
chives, 7
*chong wei zi*, 69
chrysanthemum, 41, 113
*Chrysanthemum parthenium*, 113
*Chrysanthemum vulgare*, 113
*chuan huang bai*, 84
*chuan jiao*, 124
*chuan liang zi*, 75
*chuan niu xi*, 40
*chuan xin lian*, 9
*chuan xiong*, 70
*ci wu jia*, 45
*Cibotium barometz*, 29
*Cichorium intybus*, 29
cilantro, 36
*Cimicifuga foetida*, 29
*Cimicifuga racemosa*, 29-30, 171, 185, 188
*Cinchona calisaya*, 30, 185
*Cinchona ledgeriana*, 30, 185
*Cinchona officinalis*, 30, 185
*Cinchona pubescens*, 30, 185
*Cinnamomum aromaticum*, 31
*Cinnamomum camphora*, 30, 152, 154, 185
*Cinnamomum cassia*, 31, 185
*Cinnamomum verum*, 31-32, 152, 154, 185
*Cinnamomum zeylanicum*, 31
cinnamon, 31-32, 152-153
cinquefoil, 91
*Cistanche salsa*, 32
*Citrus aurantifolia*, 32
*Citrus aurantium*, 32, 177
*Citrus bergamia*, 33
*Citrus limon*, 33
*Citrus limonia*, 33
*Citrus reticulata*, 33
clary, 102
clary sage, 102
cleavers, 54
*Clematis chinensis*, 33
clove, 112
*Cnicus benedictus*, 33-34, 185

cnidium, 34
*Cnidium monnieri*, 34
codonopsis, 34
*Codonopsis pilosula*, 34
*Codonopsis tangshen*, 34
*Coffea arabica*, 34, 172, 176, 185
coffee, 34, 83, 175, 183, 199
coin-leaved desmodium, 42
*Coix lacryma-jobi*, 34, 185
cola, 35, 172, 176, 185
*Cola acuminata*, 35, 172, 176, 185
*Cola nitida*, 35, 172, 176, 185
collinsonia, 35
*Collinsonia canadensis*, 35
coltsfoot, 117-118, 150-151, 183, 188
comfrey, 111-112, 139, 150-151, 182-183, 188
*Commiphora madagascariensis*, 35, 171, 185
*Commiphora molmol*, 35, 171, 185
*Commiphora mukul*, 35, 171, 185
*Commiphora myrrha*, 35, 171, 185
common comfrey, 111
common echinacea, 44
common elderberry, 102
common European linden, 116
common garden peony, 80
common horsetail, 47
common juniper, 65
common lavender, 68
common motherwort, 68
common periwinkle, 122
common sage, 102
common scouring rush, 48
common thyme, 116
*Convallaria majalis*, 35, 141, 189
*Conyza canadensis*, 36
*Coptis chinensis*, 36, 137-138, 171, 185
*Coptis groenlandica*, 36, 138, 185
*Cordia salicifolia*, 36
cordyceps, 36
*Cordyceps sinensis*, 36
coriander, 36
*Coriandrum sativum*, 36
corn, 125
cornflower, 25, 183
cornsilk, 125
*Cornus officinalis*, 37

corydalis, 37, 171, 183, 185
*Corydalis yanhusuo*, 37, 171, 185
*Corylus avellana*, 37
*Corylus cornuta*, 37
*Corynanthe yohimbi*, 83
costus, 104
cotton, 59, 183
cotton root bark, 59
couch grass, 45
cowslip, 91
cramp bark, 122
cranesbill, 57
*Crataegus laevigata*, 37
*Crataegus monogyna*, 37
*Crataegus pinnatifida*, 37
creosote bush, 67
*Crocus sativus*, 38, 165, 169, 171, 185
cubeb, 86
*Cullen corylifolia*, 38, 165, 177, 185
culver's-root, 69
cumin, 38
*Cuminum cyminum*, 38
*Cuphea balsamona*, 38, 186
*Curculigo orchioides*, 39, 189
*Curcuma aromatica*, 39, 171, 186
*Curcuma domestica*, 39, 171, 186
*Curcuma longa*, 39, 171, 186
*Curcuma zedoaria*, 39, 186
curled dock, 100
curly dock, 100
cuscuta, 39
*Cuscuta chinensis*, 39
*Cuscuta japonica*, 39
*Cyamopsis tetragonolobus*, 39-40, 165, 167
cyathula, 40, 183, 186
*Cyathula officinalis*, 40, 186
*Cymbopogon citratus*, 40, 169, 171, 186
cynanchum, 40
*Cynanchum atratum*, 40
cynomorium, 40
*Cynomorium songaricum*, 40
cyperus, 40
*Cyperus rotundus*, 40
*Cytisus scoparius*, 41, 56, 164-165, 189

*da huang*, 97
*da qing ye*, 64
*da zao*, 125
*Daemonorops draco*, 41
dagger plant, 124
Dalmatian sage, 102
damask rose, 99
damiana, 117
*dan shen*, 101
dandelion, 114
*dang gui*, 11
*dang shen*, 28, 34, 184
*Daucus carota*, 41, 186
dead nettle, 67
deadly nightshade, 18
*Dendranthema* x *morifolium*, 41
dendrobium, 41
*Dendrobium nobile*, 41
desert tea, 46
*Desmodium styracifolium*, 42
devil's claw, 60
devil's club, 80
devil's dung, 52
*di gu pi*, 72
digitalis, 37, 42, 88, 108, 138-141, 166, 189
*Digitalis purpurea*, 42, 141, 189
dill, 10
*Dimocarpus longan*, 42
*Dioscorea opposita*, 42
*Dioscorea villosa*, 42
*Dipsacus asper*, 43
*Dipsacus japonicus*, 43
*Dipteryx odorata*, 43, 189
*Dipteryx oppositifolia*, 43, 189
dodder, 39
dog grass, 45
dog rose, 99
dogwood fruit, 37
*dong chong zia cao*, 36
*dong gua pi*, 19
dong quai, 11, 183
dragon's blood, 41
dragon's blood palm, 41
drynaria, 43
*Drynaria fortunei*, 43
*Dryopteris filix-mas*, 43, 189
*du huo*, 10

*du zhong*, 50
dulse, 97
Dutch tonka, 43
dwarf-lilyturf root, 80
dyer's broom, 56, 183
dyer's greenwood, 56
Dyer's-woad, 64

*e zhu*, 39
East Indian sandalwood, 103
East Indian sarsaparilla, 60
eastern arborvitae, 115
eastern burningbush, 50
eastern red cedar, 66, 189
eastern white cedar, 115
echinacea, 43-44, 137, 150
*Echinacea angustifolia*, 43
*Echinacea pallida*, 44
*Echinacea purpurea*, 44
*Echinodorus macrophyllus*, 44
eclipta, 44
*Eclipta prostata*, 44
elder flower, 102
elderberry, 102
elecampane, 63-64, 183, 188
*Elettaria cardamomum*, 44
eleuthero, 45
*Eleutherococcus senticosus*, 45
*Elytrigia repens*, 45
emblic, 85
English hawthorn, 37
English lavender, 68
English plantain, 88
English tonka, 43
ephedra, 45-46, 173-176, 183, 186, 188
*Ephedra distachya*, 45-46, 174, 176, 186, 188
*Ephedra equisetina*, 45, 174, 176, 186, 188
*Ephedra gerardiana*, 45, 174, 176, 186, 188
*Ephedra intermedia*, 45, 174, 176, 186, 188
*Ephedra nevadensis*, 46
*Ephedra sinica*, 45, 174, 176, 186, 188
*Epigaea repens*, 46
*Epilobium angustifolium*, 47

*Epilobium parviflorum*, 47
epimedium, 47
*Epimedium grandiflorum*, 47
*Equisetum arvense*, 47-48
*Equisetum hyemale*, 48, 186
*Equisetum telmateia*, 47
*Eriobotrya japonica*, 48, 142-143
*Eriodictyon californicum*, 48
*Eriodictyon tomentosum*, 48
*Eryngium maritinum*, 49
*Eryngium planum*, 49
*Eryngium yuccifolium*, 49
eryngo, 49
*Erythrina indica*, 49
*Erythrina variegata*, 49
*Erythroxylum catuaba*, 49
*Eschscholzia californica*, 49, 174, 186
eucalyptus, 49, 157
*Eucalyptus globulus*, 49, 157
eucommia, 50
*Eucommia ulmoides*, 50
*Euonymus atropurpureus*, 50, 189
*Eupatorium perfoliatum*, 50
*Eupatorium purpureum*, 50, 151, 182, 186, 188
euphorbia, 51, 169
*Euphorbia pilulifera*, 51, 169
*Euphoria longan*, 42
*Euphrasia officinalis*, 51
European barberry, 19
European elder, 102
European filbert, 37
European hazel, 37
European linden, 116
European mistletoe, 123
European pennyroyal, 75, 183
European peony, 80
European sanicle, 103
European vervain, 121, 183
European wild pansy, 122
*Euryale ferox*, 51
euryale seed, 51
evening primrose, 79
*Evernia furfuracea*, 51, 159
*Evernia prunastri*, 51, 159
evodia, 52
*Evodia rutaecarpa*, 52
eyebright, 51, 150

false hellebore, 120
false sarsaparilla, 12
false unicorn root, 27
*fang feng*, 68
*fang ji*, 110
fennel, 52-53, 144
fennel seed, 53
fenugreek, 117, 183
*Ferula assa-foetida*, 52, 171, 186
*Ferula foetida*, 52, 171, 186
*Ferula rubricaulis*, 52, 171, 186
fever grass, 40
feverfew, 113, 183
field horsetail, 47
field sorrel, 100
figwort, 105
filbert leaf, 37
file, 103
*Filipendula ulmaria*, 52, 155
fireweed, 47
flax, 70, 166
florist's chrysanthemum, 41
flowering quince, 27
*Foeniculum vulgare*, 53, 144-145
forsythia, 53, 171, 183, 186
*Forsythia suspensa*, 53, 171, 186
fo-ti, 90
*Fouquieria splendens*, 53, 186
fox nut, 51
foxglove, 42, 95, 139-140
*Fragaria vesca*, 53
*Fragaria virginiana*, 53
fragrant angelica, 10
frankincense, 21
*Fraxinus americana*, 53
*Fraxinus excelsior*, 53
French lavender, 68
French psyllium, 87
French tarragon, 16
fringetree, 29
*Fritillaria cirrhosa*, 54, 186
*Fritillaria thunbergii*, 54, 186
fritillary, 54, 183
*fu ling*, 124
*fu pen zi*, 100
*fu shen*, 124
*fu zi*, 3
*Fucus vesiculosus*, 54, 146, 186, 188

*Galium aparine*, 54
*Galium odoratum*, 54
*Galium verum*, 54
gambir, 118, 157
*gan cao*, 58
*Ganoderma lucidum*, 55
*gao ben*, 70
garden heliotrope, 120
garden sage, 102
garden thyme, 116
garden valerian, 120
gardenia fruits, 55
*Gardenia jasminoides*, 55
garlic, 6-7, 183, 188
gastrodia, 55
*Gastrodia elata*, 55
*Gaultheria procumbens*, 55, 155
*ge gen*, 93
*Gelidiella acerosa*, 55, 165, 167
*Gelidium amansii*, 56, 167
*Gelidium cartilagineum*, 56, 167
*Gelidium crinale*, 56, 167
*Gelidium divaricatum*, 56, 167
*Gelidium pacificum*, 56, 167
*Gelidium vagum*, 56, 167
*Genista tinctoria*, 56, 169, 186
gentian, 56-57, 114
*Gentiana lutea*, 56
*Gentiana macrophylla*, 57
*Gentiana scabra*, 57
*Geranium maculatum*, 57, 157
German chamomile, 74
germander, 105, 115, 189
giant fennel, 52
giant horsetail, 47
ginger, 16, 125, 183
ginkgo, 57-58, 174
*Ginkgo biloba*, 57-58, 174
ginseng root, 81
*Glehnia littoralis*, 58
glehnia root, 58
glossy privet, 70
*Glycyrrhiza echinata*, 58, 186, 188
*Glycyrrhiza glabra*, 58, 186, 188
*Glycyrrhiza uralensis*, 58, 186, 188
golden bells, 53
golden eye-grass, 39
golden trumpet, 49
goldenrod, 108

## PLANT INDEX

goldenseal, 62, 137, 183
goldthread, 36, 183
goosegrass, 54
*Gossypium herbaceum*, 59, 165, 171, 186
*Gossypium hirsutum*, 59, 165, 171, 186
gotu kola, 26*gou ji*, 29
*gou qi zi*, 72
grains of paradise, 9
grass-leaved calamus, 4
grassy-leaved sweetflag, 4
gravel root, 50, 151
great blue lobelia, 71
greater celandine, 28
greater galangal, 8, 66
Grecian laurel, 68
green peppercorns, 87
green tea, 22
grifola, 59
*Grifola frondosa*, 59
*Grifola umbellata*, 59
grindelia, 59
*Grindelia robusta*, 59
*Grindelia squarrosa*, 59
*gu sui bu*, 43
*gua lou*, 116
*gua lou ren*, 116
*guang jing qian cao*, 42
guar gum, 39-40, 165
guarana, 83
guelder rose, 122
guggul, 35, 183
*gui zhi*, 31
gumweed, 59

*hai tong pi*, 49
*Hamamelis virginiana*, 59-60, 156-157
*han fang ji*, 110
*han lian cao*, 44
hardy rubber tree, 50
*Harpagophytum procumbens*, 60
*he huan pi*, 5
*he shou wu*, 90
*he ye*, 78
*he zi*, 114
heal-all, 91
healing-herb, 111
heartsease, 122

heather, 22
*Hedeoma pulegioides*, 60, 164, 186
*Helianthus annuus*, 60
hemidesmus, 60
*Hemidesmus indicus*, 60
hen of the woods, 59
henna, 68, 182
*Hepatica nobilis*, 61, 186
herb-of-grace, 101
Hercules' club, 124
*Heuchera micrantha*, 61, 157
hibiscus, 61
*Hibiscus sabdariffa*, 61
high mallow, 73
Himalayan mayapple, 88, 189
*ho shou wu*, 90
hoary plantain, 88
hollyhock, 5
holy thistle, 33
*hong hua*, 24
hops, 61
*Hordeum vulgare*, 61, 186
horehound, 18, 72, 74, 183
horseheal, 64
horsemint, 76
horseradish, 14
horsetail, 47-48
*hou po*, 72
*hou po hua*, 72
*hu lu ba*, 117
*huang bai*, 84
*huang jing*, 90
*huang lian*, 36
*huang qi*, 17
*huang qin*, 105
*huangpishu*, 84
huckleberry, 119
*Humulus lupulus*, 61
Hungarian chamomile, 74
*huo xiang*, 5, 89
hydrangea, 62, 143
*Hydrangea arborescens*, 62, 143
*Hydrastis canadensis*, 62, 137-138, 171, 186
*Hypericum perforatum*, 62, 157, 173-174, 177
hyssop, 63, 183
*Hyssopus officinalis*, 63, 171, 186

Iceland moss, 26
*Ilex paraguayensis*, 63, 157, 176
*Illicium verum*, 63
*Imperata cylindrica*, 63
Indian almond, 114
Indian jointfir, 45
Indian physic, 12
Indian poke, 120
Indian sarsaparilla, 60
Indian senna, 106
Indian tobacco, 71
Indian valerian, 120
inmortal, 17, 183
*Inula britannica*, 63
*Inula helenium*, 64, 186, 188
ipe roxo, 113
ipecac, 12, 26, 167-168, 183
*Ipomoea purga*, 64, 189
*Iris germanica*, 64
*Iris pallida*, 64
*Iris versicolor*, 64, 169, 186
*Iris virginica*, 64, 169, 186
*Isatis tinctoria*, 64
ispaghula, 87

jaborandi, 85-88, 186, 189-190
Jack-in-the-pulpit, 13
jalap, 64, 189
Jamaican quassia, 85
jambolan, 113
Japanese apricot, 92
Japanese arisaema, 13, 183
Japanese cornel, 37
Japanese honeysuckle, 72
Japanese teasel, 43
jasmine, 65
*Jasminum grandiflorum*, 65
jatamansi, 78, 171, 183, 186
*jiang huang*, 39
*jie geng*, 88
*jin yin hua*, 72
Job's tears, 34, 183
Joe Pye, 50, 182-183, 188
Joe-Pye-weed, 50
johimbe, 83
Johnny jump up, 122
Joshua tree, 124
*ju hua*, 41

*Juglans cinerea*, 65
*Juglans nigra*, 65
jujube, 125, 183
jujube seeds, 125, 183
jumbul, 113
juniper, 65-66, 184
*Juniperus communis*, 65, 186
*Juniperus monosperma*, 66, 186
*Juniperus osteosperma*, 66, 186
*Juniperus oxycedrus*, 65, 186
*Juniperus virginiana*, 66, 186, 189


*Kaempferia galanga*, 66
Kansas snakeroot, 43-44
kava, 86-87
kava pepper, 86
kava-kava, 86
kelp, 79, 146
kinnikinnik, 13
kola nut, 35, 175
Korean ginseng, 81
Krameria, 66, 157
*Krameria argentea*, 66, 157
*Krameria triandra*, 66, 157
*ku lian pi*, 74-75
*ku shen*, 109
kudzu, 93


*Lactuca quercina*, 66
*Lactuca serriola*, 66
*Lactuca virosa*, 66-67
lady's bedstraw, 54
lady's mantle, 5
*Lamium album*, 67
lapacho colorado, 113
lapacho morado, 113
large-leaf gentian, 57
large leafed linden, 116
*Larrea tridentata*, 67
*Laurus nobilis*, 68
*Lavandula angustifolia*, 68
*Lavandula latifolia*, 68
*Lavandula stoechas*, 68
*Lavandula x intermedia*, 68
lavender, 68
*Lawsonia inermis*, 68, 182
*Ledebouriella seseloides*, 68

## PLANT INDEX

lemandarin, 33
lemon, 8, 33, 75, 115
lemon balm, 75
lemon thyme, 115
lemon verbena, 8
lemongrass, 40, 169, 184
*Lentinus edodes*, 68
*Leonurus artemisia*, 69
*Leonurus cardiaca*, 68, 171, 186
*Leonurus heterophyllus*, 69, 171, 186
*Leonurus sibiricus*, 69, 170-171, 186
*Leptandra virginica*, 69, 186
lesser galangal, 8
levant cotton, 59
*Levisticum officinale*, 69, 171, 186
lian fang, 78
lian qiao, 53
lian xu, 78
lian zi, 78
lian zi xin, 78
licorice, 12, 58, 184, 188
*Ligusticum chuanxiong*, 70, 186
*Ligusticum porteri*, 70, 186
*Ligusticum sinense*, 70
*Ligusticum wallichii*, 70
ligusticum, 70, 186
ligustrum fruits, 70
*Ligustrum lucidum*, 70
*Lilium brownii*, 70
lily of the valley, 35, 189
lime, 32-33, 116
lime leaves, 116
lime tree flower, 116
linden, 116
*ling chih*, 55
*ling chih* mushroom, 55
*ling zhi*, 55
linseed, 70
*Linum usitatissimum*, 70, 166-167
*Lippia citriodora*, 8
liquorice, 58
liverwort herb, 61
lobelia, 71, 168-169, 183-184, 186
*Lobelia inflata*, 71, 168-169, 186
*Lobelia siphilitica*, 71, 186
lomatium, 71, 184, 186
*Lomatium dissectum*, 71, 186
long birthwort, 13, 189
*long dan*, 57

*long yan rou*, 42
longan, 42
*Lonicera japonica*, 72
loquat, 48, 142
lotus, 78, 199
lovage, 69-70, 184
lungwort, 94
lycium, 72, 184, 186
*Lycium barbarum*, 72
*Lycium chinense*, 72, 186
*Lycopus americanus*, 72, 186, 188
*Lycopus europaeus*, 72 186, 188
*Lycopus virginicus*, 72, 186, 188

ma dou ling, 14, 189
ma huang, 45
mace, 77, 173, 184
Madagascar periwinkle, 24, 189
Madagascar vanilla, 120
magnolia, 72-73, 104, 131, 184, 186
*Magnolia liliflora*, 72
*Magnolia officinalis*, 72, 131, 186
magnolia vine, 104
*Magnolia virginiana*, 73
*Mahonia aquifolium*, 73, 138, 186
*Mahonia nervosa*, 73, 138, 186
*Mahonia repens*, 73, 138, 186
mai dong, 80
mai ya, 61
maidenhair fern, 4, 184
maidenhair tree, 57
maitake, 59
male fern, 43, 182, 184, 188
malva, 73
*Malva sylvestris*, 73
mandarin lime, 33
mandarin orange, 33
*Mandragora officinarum*, 73, 133-134, 189
mandrake, 73, 89, 133, 189
mao gen, 63
*Maranta arundinacea*, 74
*Marrubium vulgare*, 74, 171, 186
marshmallow, 9
maté, 63, 175
*Matricaria recutita*, 74
mayapple, 88-89, 189
maypop, 82

meadowsweet, 52
mealy kudzu, 93
*Medicago sativa*, 74
*mei gui hua*, 99
*Melia azedarach*, 74, 169, 189
*Melia toosendan*, 75, 189
*Melissa officinalis*, 75
*Mentha piperita*, 75
*Mentha pulegium*, 60, 75, 164, 171, 186
*Mentha spicata*, 76
*Menyanthes trifoliata*, 76
Mexican valerian, 120
Mexican vanilla, 120
Mexican yam, 42
microcos leaves, 76
*Microcos nervosa*, 76
milfoil, 3
milk thistle, 107
milkweed, 12, 17
mimosa tree, 5
*ming dang shen*, 28, 184
Missouri snakeroot, 82
*Mitchella repens*, 76
*Monarda clinopodia*, 76, 171, 186
*Monarda didyma*, 76, 171, 186
*Monarda fistulosa*, 76, 171, 186
*Monarda pectinata*, 76, 171, 186
*Monarda punctata*, 76, 171, 186
morinda, 77
*Morinda officinalis*, 77
Mormon tea, 46
*Morus alba*, 77
motherwort, 68-69, 170, 183-184
*mu dan pi*, 80
*mu gua*, 27
*mu xiang*, 14, 104, 189
*mu zei*, 48
mugwort, 16, 184
muira puama, 93
mullein, 121
mum, 41
muscatel sage, 102
mustard, 21-22, 108
*Myrcia sphaerocarpa*, 77
*Myrica cerifera*, 77, 157
*Myrica pensylvanica*, 77, 157
*Myristica fragrans*, 77, 152, 154, 173-174, 186, 190

*Myroxylon balsamum*, 78
myrrh, 35, 184

*nan sha shen*, 4
nard, 78
*Nardostachys jatamansi*, 78, 171, 186
narrow-leaved echinacea, 43
narrow-leaved purple coneflower, 43
*Nasturtium officinale*, 78, 171, 186
*Nelumbo nucifera*, 78
*Nepeta cataria*, 79, 171, 186
*Nereocystis luetkeana*, 79, 146
nettles, 119
New Jersey tea, 25
night-blooming cereus, 105
northern prickly ash, 124
northern white cedar, 115
notopterygium, 79
*Notopterygium incisum*, 79
*nu zhen zi*, 70
*nui xi*, 3
nut grass, 40
nutmeg, 77, 152-153, 173, 184

oak, 51, 94, 156-157
oak moss, 51
oats, 18
*Ocimum basilicum*, 79, 144-145, 154
ocotillo, 53, 184
*Oenothera biennis*, 79
old man's beard, 119
olibanum, 21
oneseed hawthorn, 37
oneseed juniper, 66
*Ophiopogon japonicus*, 80
opium poppy, 82
*Oplopanax horridus*, 80
opobalsam, 78
orangeroot, 62
oregano, 80
Oregon barberry, 73
Oregon grape, 73, 184
Oregon grapeholly, 73
oriental arborvitae, 88
oriental ginseng, 81
*Origanum majorana*, 80
*Origanum vulgare*, 80

## PLANT INDEX

orris, 64
osha, 70, 184
Oswego tea, 76
*ou jie*, 78
our Lady's bedstraw, 54
our Lord's candle, 124

Pacific valerian, 120
*Paeonia lactiflora*, 80
*Paeonia officinalis*, 80
*Paeonia suffruticosa*, 80, 186
pale purple coneflower, 44
pale-flowered echinacea, 44
palma christi, 98
*Panax ginseng*, 81
*Panax notoginseng*, 82, 186
*Panax quinquefolius*, 82
*Papaver somniferum*, 82
papaya, 24
para' cress, 109
Paraguay cress, 109
Paraguay tea, 63
Paraguayan sweet herb, 110
*Parietaria judaica*, 82
*Parietaria officinalis*, 82
parsley, 36, 84, 184
*Parthenium integrifolium*, 43-44, 82
partridge berry, 76
*Passiflora incarnata*, 82
*Passiflora laurifolia*, 83
passion flower, 82-83
passion flower herb, 83
*pata de vaca*, 19
patchouly, 89
pau d'arco, 113
*Paullinia cupana*, 83, 157, 176
*Pausinystalia yohimbe*, 83, 173-174
peach, 92, 141-142, 184, 189
pearly everlasting, 9
pedra hume caa, 77
*Pelargonium graveolens*, 83
pellitory of the wall, 82
pennyroyal, 60, 75, 163-164, 183, 199
peony, 80, 184
peony root, 80
pepper, 86-87, 104, 124, 184
peppermint, 75
periwinkle, 24, 122, 189

Peruvian rhatany, 66
*Petroselinum crispum*, 84, 171, 186
*Peumus boldus*, 84
pfaffia, 84
*Pfaffia paniculata*, 84
*Phellodendron amurense*, 84, 137-138, 18
phellodendron bark, 84, 184
*Phellodendron chinense*, 84, 138, 186
*Phoradendron leucarpum*, 85, 190
*Phyllanthus emblica*, 85
*Phytolacca americana*, 85, 147-148, 190
*pi pa ye*, 48
*Picrasma excelsa*, 85, 186
pill-bearing spurge, 51
*Pilocarpus jaborandi*, 85-88, 186, 190
*Pilocarpus microphyllus*, 85, 186, 190
*Pilocarpus pennatifolius*, 85, 186, 190
*Pimenta dioica*, 86
*Pimpinella anisum*, 86, 186
pineapple weed, 27
pinellia, 86, 184, 186
*Pinellia ternata*, 86, 186
piney, 80
pink pepper, 104
pink peppercorns, 104
pinkroot, 109
*Pinus strobus*, 86
*Piper cubeba*, 86
*Piper methysticum*, 86
*Piper nigrum*, 87
pipsissewa, 28
*Plantago asiatica*, 87, 187
*Plantago lanceolata*, 88
*Plantago major*, 88
*Plantago media*, 88
*Plantago ovata*, 87, 166-167
plantain, 88
*Platycladus orientalis*, 88, 159
*Platycodon grandiflorum*, 88
pleurisy root, 17, 184
*Podophyllum emodi*, 88
*Podophyllum hexandrum*, 88, 186, 190
*Podophyllum peltatum*, 8, 186, 190
*Pogostemon cablin*, 89
poke, 85, 120, 147, 189
polygala, 89, 171, 186
*Polygala senega*, 89, 171, 186
*Polygala sibirica*, 89

*Polygala tenuifolia*, 89
*Polygonatum biflorum*, 89
*Polygonatum odoratum*, 89
*Polygonatum sibiricum*, 90
*Polygonum bistorta*, 90
*Polygonum multiflorum*, 90
polyporus, 59, 124
*Polyporus umbellatus*, 59
pomegranate, 94, 189
poplar buds, 90
poppyseed, 82
*Populus balsamifera*, 90, 155
*Populus x jackii*, 90, 155
*Portulaca oleracea*, 90, 149, 186
pot marigold, 22
*Potentilla erecta*, 91, 157
prairie dock, 82
prickly ash, 124, 171, 184
*Primula veris*, 91
prince's pine, 28
prinsepia, 91
*Prinsepia uniflora*, 91
provence rose, 99
*Prunella vulgaris*, 91
*Prunus armeniaca*, 91, 143, 190
*Prunus dulcis*, 92, 143, 190
*Prunus mume*, 92
*Prunus persica*, 92, 142-143, 187, 190
*Prunus serotina*, 92, 142-143
*Prunus spinosa*, 93, 143
psoralea, 38, 177, 184
*Psoralea corylifolia*, 38
psyllium, 87, 165-166
*Pterocarpus santalinus*, 93
*Ptychopetalum olacoides*, 93
*Ptychopetalum uncinatum*, 93
pubescent angelica, 10
*Pueraria lobata*, 93
*Pueraria thomsonii*, 93
puke weed, 71
*Pulmonaria officinalis*, 94
*Punica granatum*, 94, 157, 190
purging buckthorn, 95, 184, 188
purple coneflower, 43-44
purslane, 90, 184

qian shi, 51
qiang huo, 79
qin jiao, 57
qing guo, 23
qing hao, 15
qing mu xiang, 14, 189
qing pi, 33
quassia, 85, 94, 184, 187
*Quassia amara*, 94, 187
Queen Ann's lace, 41
queen-of-the-meadow, 50, 52
queen's root, 111
queen's-delight, 111
*Quercus alba*, 94, 157
*Quercus petraea*, 94, 157
*Quercus robur*, 94, 157
quillaja, 95, 149, 157, 172
*Quillaja saponaria*, 95, 149, 157, 172
quinine, 30, 184

raspberry, 99
red cinchona, 30
red clover, 117, 184
red magnolia, 72
red puccoon, 103
red raspberry, 99
red root, 103
red sandalwood, 93
red saunders, 93
red sorrel, 61
red-rooted sage, 101
rehmannia, 95
*Rehmannia glutinosa*, 95
reishi, 55
ren dong teng, 72
ren shen, 81-82
*Rhamnus catharticus*, 95-96, 178-179, 187-188
*Rhamnus frangula*, 95, 178-179, 187-188
*Rhamnus purshiana*, 96, 178-179, 187-188
rhatany, 66, 157
*Rheum officinale*, 97, 149, 157, 179, 187-188
*Rheum palmatum*, 97, 149, 157, 179, 187-188
*Rheum tanguticum*, 97, 149, 157, 179, 187-188
*Rhodymenia palmetta*, 97, 146

# PLANT INDEX

rhubarb, 97, 178, 183, 188
*Rhus coriaria*, 98, 157
*Rhus glabra*, 98, 157
*Ribes nigrum*, 98
richweed, 35
*Ricinus communis*, 98, 147-148, 178-179, 187
rockwrack, 54
Roman chamomile, 27, 169, 184
*Rosa alba*, 99
*Rosa canina*, 99
*Rosa centifolia*, 99
*Rosa damascena*, 99
*Rosa gallica*, 99
*Rosa rugosa*, 99
rose geranium, 83
roselle, 61
rosemary, 99, 169, 184
*Rosmarinus officinalis*, 99, 165, 169, 171, 187
*rou cong rong*, 32
*rou gui*, 31
rough horsetail, 48
royal jasmine, 65
*Rubus chingii*, 100
*Rubus fruticosus*, 99
*Rubus idaeus*, 99
*Rubus strigosus*, 99
*Rubus suavissimus*, 100
rue, 101, 184
rugose rose, 99
*rui ren*, 91
*Rumex acetosa*, 100, 149, 157
*Rumex acetosella*, 100, 149, 157
*Rumex crispus*, 100, 149, 157
*Rumex crispus*, 100, 149, 157
*Rumex hymenosepalus*, 100, 157
*Rumex obtusifolius*, 100, 149, 157
*Ruscus aculeatus*, 100
Russian belladonna, 105
Russian comfrey, 111, 150, 182
*Ruta graveolens*, 101, 165, 171-172, 177, 187

sabal, 107
safflower, 24, 170-171, 184
saffron, 38, 169, 184
sage, 101-102, 115, 158, 184
*Salix alba*, 101, 155, 157
*Salvia columbariae*, 101
*Salvia hispanica*, 101
*Salvia miltiorrhiza*, 101
*Salvia officinalis*, 102, 159, 187
salvia root, 101
*Salvia sclarea*, 102
*Sambucus canadensis*, 102
*Sambucus nigra*, 102
*san qi ginseng*, 82, 184
sanchi ginseng, 82
sandalwood, 93, 103
sandalwood Padauk, 93
*sang bai pi*, 77
*sang shen*, 77
*sang ye*, 77
*sang zhi*, 77
*Sanguinaria canadensis*, 103, 137-138, 169, 172, 187
sanicle, 103
*Sanicula europaea*, 103
*Santalum album*, 103
sarsaparilla, 12, 60, 108
sassafras, 103-104, 152-154
*Sassafras albidum*, 103-104, 154
*Satureja hortensis*, 104
*Satureja montana*, 104
*Saussurea lappa*, 104
saw palmetto, 107
sawbrier, 108
scabrous gentian, 57
scabwort, 64
*Schinus molle*, 104, 172
*Schinus terebinthifolia*, 104, 172
*Schisandra chinensis*, 104
schizandra, 104
scopolia, 105, 189-190
*Scopolia carniolica*, 105, 190
Scotch broom, 41, 189
scouring rush, 48, 184
*Scrophularia marilandica*, 105
*Scrophularia nodosa*, 105
scullcap, 105
scurfy pea, 38
scute, 105
*Scutellaria baicalensis*, 105
*Scutellaria lateriflora*, 105
scythian lamb, 29
sea holly, 49

*Selenicereus grandiflorus*, 105-106
self-heal, 91
Seneca snakeroot, 89, 184
Senega snakeroot, 89
senna, 106-107, 109, 178-179, 184, 187-188
*Senna alexandrina*, 106, 179, 187
*Senna obtusifolia*, 106, 179, 187-188
*Senna tora*, 106, 179, 187-188
*Serenoa repens*, 107
serpentaria, 14, 132, 185, 189
sesame, 107
*Sesamum orientale*, 107
sete sangrias, 38, 184
Seville orange, 32
Seychelles cinnamon, 31
*shan yao*, 42
*shan zha*, 37
*shan zhu yu*, 37
shave grass, 47
shavetail grass, 47
*she chuang zi*, 34
sheep sorrel, 100
*sheng di huang*, 95
*sheng ma*, 29
shepherd's purse, 23, 184
*shi chang pu*, 4
*shi hu*, 41
*shi liu gen pi*, 94
*shi liu pi*, 94
shiitake, 68
shiitake mushroom, 68
shrubby sophora, 109
Siberian ginseng, 45
Siberian Solomon's-seal, 90
Sichuan lovage, 70, 184
Sichuan pagoda tree, 75, 189
Sichuan pepper, 124, 184
siler, 68
silk tree, 5, 170, 184
*Silybum marianum*, 107
*Sinapis alba*, 108
skullcap, 105
skunk cabbage, 112, 172
slippery elm, 118
sloe, 93
small spikenard, 12, 184
small-flowered willow herb, 47
*Smilax febrifuga*, 108

*Smilax glauca*, 108
*Smilax medica*, 108
*Smilax ornata*, 108
*Smilax regelii*, 108
soap tree, 95
soapweed, 124
*Solidago canadensis*, 108
*Solidago gigantea*, 108
*Solidago virgaurea*, 108
Solomon's seal, 89
*Sophora flavescens*, 109
sorrel, 61, 100
sour orange, 32
southern blue gum, 49
southern prickly ash, 124
southern tsangshu, 18
southernwood, 15, 184
Spanish bayonet, 124
Spanish chestnut, 24
Spanish jasmine, 65
Spanish lavender, 68
Spanish saffron, 38
spearmint, 76
speedwell, 121
*Spigelia marilandica*, 109
spignet, 12
spike lavender, 68
spikenard, 12, 78, 183-184
*Spilanthes acmella*, 109
*Spilanthes oleracea*, 109
*Spiraea ulmaria*, 52
spotted beebalm, 76
spreading dogbane, 12, 189
squawvine, 76
St. John's wort, 62
*Stachys officinalis*, 109
star anise, 63
stargrass, 5
starwort, 5, 64
*Stellaria media*, 110
stephania root, 110
*Stephania tetrandra*, 110
stevia, 110
*Stevia rebaudiana*, 110
stillingia, 111, 188
*Stillingia sylvatica*, 111, 188
stinging nettle, 119
stoneroot, 35
strawberry, 53

PLANT INDEX 229

*Styrax benzoin*, 111
*Styrax paralleloneurum*, 111
*Styrax tonkinensis*, 111
*suan zao ren*, 125
suma, 84
sumac, 98
Sumatra benzoin, 111
summer savory, 104
sunflower, 60
*suo yang*, 40
Surinam quassia, 94
swamp cedar, 115
sweet annie, 15, 184
sweet basil, 79
sweet bay, 68
sweet birch, 20, 155
sweet flag, 3, 64
sweet marjoram, 80
sweet tea, 100
sweet violet, 122
sweet woodruff, 54
sweetbay, 73
sweetleaf, 110
*Symphytum asperum*, 111, 151, 182
*Symphytum officinale*, 111-112, 151, 182, 187-188
*Symphytum* x *uplandicum*, 111, 151, 182
*Symplocarpus foetidus*, 112, 149, 172
*Syzygium aromaticum*, 112
*Syzygium cumini*, 113
Szechuan lovage, 70
Szechuan pepper, 124
Szechuan teasel, 43


*Tabebuia heptaphylla*, 113
*Tabebuia impetiginosa*, 113
tacamahac, 90
taheebo, 113
Tahitian vanilla, 120
*tan xiang*, 103
*Tanacetum parthenium*, 113, 187
*Tanacetum vulgare*, 113, 159, 164-165, 171, 187, 190
*tang kuei*, 11
tangerine, 33
tansy, 113, 164, 189
*tao ren*, 92

*Taraxacum officinale*, 114
Tasmanian blue gum, 49
tea, 8, 15, 22, 25, 29, 31-33, 39, 46, 48-49, 52, 54-56, 58, 63, 74, 76, 81, 85, 88, 93, 95-98, 100, 118, 123, 144, 156-158, 169, 175
tea berry, 55
teasel, 43
*Terminalia bellerica*, 114, 157
*Terminalia chebula*, 114, 157
*Ternstroemia pringlei*, 114
*Teucrium chamaedrys*, 115, 190
*Teucrium scorodonia*, 115
thoroughwort, 50
thuja, 88, 115, 159, 165, 171, 184, 187
*Thuja occidentalis*, 115, 159, 165, 171, 187
*Thuja orientalis*, 88
thyme, 115-116
*Thymus vulgaris*, 116
*Thymus* x *citriodorus*, 115
*tian cha*, 100
*tian hua fen*, 116
*tian ma*, 55
*tian men dong*, 17
*tian nan xing*, 13
*tian xian teng*, 14, 189
tienchi, 82
tienchi ginseng, 82
Tilia estrella, 114
*Tilia platyphyllos*, 116
*Tilia* x *europaea*, 116
tolu, 78
tonka, 43, 189
tonka bean, 43
toothache plant, 109
toothache tree, 124
tormentil, 91
trailing arbutus, 46
tree moss, 51, 119
tree peony, 80, 184
tree peony bark, 80, 184
trichosanthes, 116, 164-165, 184, 187
*Trichosanthes kirilowii*, 116, 164-165, 187
*Trifolium pratense*, 117, 187
*Trigonella foenum-graecum*, 117, 187
*Trillium erectum*, 117, 171-172, 187
triticum, 45

tropical almond, 114
true chamomile, 74
true saffron, 38
true unicorn, 5
*tu huo xiang*, 5
*tu si zi*, 39
turkey rhubarb, 97
turmeric, 39, 169, 184
*Turnera diffusa*, 117
*Tussilago farfara*, 117-118, 150-151, 187-188
twitch grass, 45

*Ulmus rubra*, 118
*Uncaria gambir*, 118, 157
*Uncaria tomentosa*, 118
*Urtica dioica*, 119
usnea, 119
*Usnea barbata*, 119
usnea lichen, 119
Ussurian thorny pepperbush, 45
uva-ursi, 13, 22, 157, 172, 184

*Vaccinium angustifolium*, 119
*Vaccinium corymbosum*, 119
*Vaccinium myrtillus*, 119
*Vaccinium pallidum*, 119
valerian, 120
*Valeriana edulis*, 120
*Valeriana officinalis*, 120
*Valeriana sitchensis*, 120
*Valeriana wallichii*, 120
vanilla, 120
*Vanilla planifolia*, 120
*Vanilla tahitensis*, 120
*Veratrum viride*, 120, 190
*Verbascum thapsus*, 121
*Verbascum densiflorum*, 121
*Verbascum phlomoides*, 121
*Verbena hastata*, 121, 187
*Verbena officinalis*, 121, 187
*Veronica officinalis*, 121
vetiver, 121, 184
*Vetiveria zizanoides*, 121, 165, 171, 187
*Viola tricolor*, 122
*Viola odorata*, 122

*Viola sororia*, 122
violet, 122
Virginia snakeroot, 14, 132, 189
Virginian strawberry, 53
*Viscum album*, 123, 147-148
*Vitex agnus-castus*, 123, 171, 187
*Viburnum opulus*, 122
*Viburnum prunifolium*, 122, 149
*Vinca minor*, 122

wahoo, 50, 189
wakerobin, 117
wallflower, 12
water horehound, 72
watercress, 78, 184
wax gourd, 19
wax myrtle, 77
*wei ling xian*, 33
West Indian lemongrass, 40
white ash, 53
white birch, 20
white Chinese olive, 23
white horehound, 74
white mulberry, 77
white nettle, 67
white oak bark, 94
white peony, 80
white pepper, 87
white pine, 86
white rose, 99
white sandalwood, 103
white saunders, 103
white willow, 101
whortleberry, 119
wild bergamot, 76
wild black cherry, 92
wild carrot, 41, 184
wild cherry, 92-93, 142
wild gentian, 56
wild geranium, 57
wild ginger, 16
wild hydrangea, 62
wild indigo, 18, 184
wild ipecac, 12
wild lettuce, 66
wild licorice, 12
wild oat, 18
wild passion flower, 82

# PLANT INDEX

wild yam, 42
willow herb, 47
winter savory, 104
wintergreen, 55, 155
witch-hazel, 59-60
*Withania somnifera*, 124, 165, 187
woad, 64
*Wolfiporia cocos*, 124
wood betony, 109
wood sage, 115
woodruff, 54
woolly blue violet, 122
woolly grass, 63
woolly yerba santa, 48
wormwood, 15, 158, 184, 188
*wu mei*, 92
*wu wei zi*, 104
*wu zhu yu*, 52

*xian mao*, 39
*xiang fu*, 40
*xin yi*, 72
*xing ren*, 91
*xu duan*, 43
*xuan fu hua*, 63
*xue jie*, 41

*yan hu suo*, 37
yarrow, 3, 171, 184
yaw root, 111
yellow bedstraw, 54
yellow cinchona, 30
yellow dock, 100
yellow gentian, 56
yellow granadilla, 83
yellow mustard, 108

yellow puccoon, 62
yellow quinine, 30
yellow sandalwood, 103
yellow saunders, 103
yellow starwort, 64
yerba mansa, 10
yerba maté, 63, 175
yerba santa, 48
*yi mu cao*, 69
*yi yi ren*, 34
*yin chen hao*, 15
*yin yang huo*, 47
ylang ylang, 23
yohimbe, 83, 173-175
*yu zhu*, 89
*yuan zhi*, 89
yucca, 124
*Yucca aloifolia*, 124
*Yucca brevifolia*, 124
*Yucca glauca*, 124
*Yucca whipplei*, 124

*Zanthoxylum americanum*, 124, 187
*Zanthoxylum clava-herculis*, 124, 187
*Zanthoxylum simulans*, 124, 187
*Zanthoxylum bungeanum*, 124, 187
*Zanthoxylum schinifolium*, 124, 187
*Zea mays*, 125
*ze xie*, 6
zedoary, 39, 184
*zhi mu*, 9
*zhi zi*, 55
*zhu ling*, 59
*Zingiber officinale*, 125, 187
*Ziziphus jujuba*, 125
*Ziziphus spinosa*, 125, 171, 187